FARM MACHINERY

Selection, Investment and Management

Resource Management Series

FARM MACHINERY
Selection, Investment and Management

Andrew Landers

FARMING
PRESS

I am grateful to the following for permission to reproduce copyright material:

John Cathie for Tables 4.3, 4.5, 4.6 and Figure 4.1
Larry Miller for Figure 6.1
Michael Barrett for Figure 12.4 and Table 12.4
Brian Finney for Tables 6.3 and 6.4
Anthony Collier for Figure 3.5
SAC for Tables 4.1, 4.4 and 6.8
John Nix for Tables 6.1, 7.1 and 7.3
Mats Bohm for Figures 8.1 and 8.2

First published 2000
Copyright © Andrew Landers

1 3 5 7 9 10 8 6 4 2

ISBN 0 85236 540 3

A catalogue record for this book is available from the British Library.

Published by Farming Press, United Business Media, Sovereign House, Sovereign Way, Tonbridge, Kent TN9 1RW, United Kingdom

Distributed in North America by Diamond Farm Enterprises, Box 537, Bailey Settlement Road, Alexandria Bay, NY 13607, USA.

Printed by Antony Rowe Ltd, Chippenham, Wiltshire

CONTENTS

FOREWORD

'That part of the holding of a farmer
or landowner which pays best for
cultivation is the small estate within
the ring fence of his skull.'

Charles Dickens, *Farm and College*, 1868

Agriculture throughout the world is undergoing great change. Increases
in production since the Second World War, in part due to improvements
developed by engineers, scientists and educators, coupled with
government intervention, have resulted in surpluses in many countries.
Governments no longer see it as important to support farming, so
farmers now have to compete in a global market. The days of farming
as a way of life are rapidly disappearing and farmers must be business
people first, farmers second.

Modern farmers are well versed in preparing financial statements,
forward planning and making key management decisions. As
commodity prices continue to fall or level out, there is a great need for
cost control. Variable costs have been tightened over the past decade, so
now is the time to consider machinery costs.

This book has been written to help farmers make important decisions
regarding the selection and management of farm machinery. On many
farms, machinery investment has been carried out on an *ad hoc* basis
but modern business management requires attention to detail in all
sectors and machinery management is no exception. This book provides
a mixture of management, engineering and science coupled with many
good practical examples for mechanisation in the twenty-first century.

Andrew Landers
Cornell University
Ithaca, New York

July 2000

To my family, Sarah, Charlotte and Rebecca and to the memory of my late father, Colin, a true gentleman.

To my former students, farmers, friends and colleagues who inspired me to write this book whilst maintaining common sense and good humour.

Chapter 1

INTRODUCTION TO MECHANISATION

The productivity of farms was increased by the introduction of tractors and their matching implements by pioneers such as Ferguson, Ford and Deere. The advent of the tractor and implement combinations led to the first requirements for machinery management.

Farm production continues to increase as engineers continue to develop new equipment and scientists develop new crop varieties. The use of satellites to give variable-rate application of inputs and monitoring of crop yields exemplifies the continuous improvement in farm machinery.

The level of mechanisation on a farm depends on many factors but a major determinant must be the question of return on investment. Will the investment in a machine give a good return on capital? Will it allow production to continue or improve and so benefit the business?

THE BENEFITS OF MECHANISATION

Improved working conditions
The major advantage of farm mechanisation is improved working conditions. People with many years' experience will recall the drudgery associated with farming operations such as hand-hoeing sugar beet, lifting potatoes by hand, lifting two-hundred-weight sacks of corn and numerous other tiring, back-breaking tasks. Farm operations have progressed, via mechanisation, in line with other industries and the resultant social benefits of modern work conditions can be found on many farms. Automation in the milking parlour and vegetable packhouse are good examples of such developments.

Increased returns
Mechanisation and good management can result in better timeliness in field operations and on good soils this can result in improved yields. Good cultivations can result in early sowing of crops, earlier ripening and better yields. The farmer who is first to market with an early, quality product can reap greater rewards.

Productivity
The average adult can produce 0.15kW-h (kilowatt-hour) of energy/h whilst working continuously, approximately the same energy value as 0.05l/h (litres/hour) of diesel fuel. A good tractor can produce 3kW-h of energy per litre of diesel. As the drift of labour from the land continues and farms grow bigger, so the area farmed by those remaining increases.

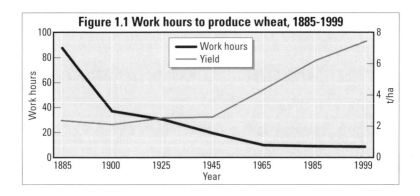

Figure 1.1 Work hours to produce wheat, 1885-1999

On certain large-scale cereal farms it is common to find one person operating equipment for every 300ha farmed (see Figure 1.1).

Quality

The advent of mechanised field-scale vegetables, well-engineered packhouse equipment and the cool-chain method of food distribution has resulted in quality produce being presented on the supermarket shelf.

Cost reduction

Economies of scale, greater output per person or per hectare and reduced costs of production are phrases often heard at farm management meetings. Attention to detail is the key and farming is no exception. The top ten per cent of farmers would be extremely good business managers irrespective of the industry in which they worked. Good mechanisation management allows successful managers the opportunity to identify, monitor and reduce their mechanisation costs (see Table 1.1).

Table 1.1 VARIATION IN POWER COSTS (based on a survey of 350ha mixed dairy/arable farms)			
Power costs	Cost £/ha		
	Average	Top 10%	Bottom 10%
Machinery repairs	48	35	58
Finance charge	25	5	45
Fuel & Electricity	57	44	58
Contractors	16	8	15
Road tax and insurance	8	6	12
Depreciation	95	64	116
Total	249	162	304

WHY IT IS IMPORTANT TO PLAN MECHANISATION

Lower returns/falling profits

After land purchase, machinery and labour are often the next most important cost on a farm. Good mechanisation management will ensure that the business facts are available to aid the decision-making process during both good and bad times.

Labour organisation

As the decline in the agricultural workforce continues, there is a growing need to consider alternatives to replacing farm staff. Improving the standard or level of mechanisation is one of the important management decisions to be made when an employee retires or leaves the business.

Capital cost of equipment

The price of farm equipment continues to rise and the increasing cost of manufacturing and distribution is passed on to the customer. Technological advances, such as precision farming equipment, require large amounts of research and development and this is reflected in the purchase price.

Crop planning

Changes in commodity prices mean that the good farmer must always look at alternative crops to produce and sell. Whilst certain crops such as oilseed rape and linseed may be sown and harvested with the same equipment as is used for cereals, other crops, such as rowcrops, require an investment in alternative machinery.

Minimised fixed costs

As incomes fall, so profit is reduced, unless good management policies are adopted. Improving output by matching machinery to resources is

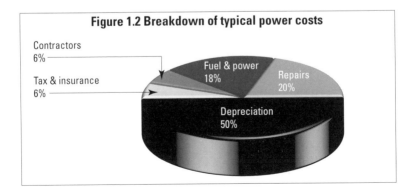

Figure 1.2 Breakdown of typical power costs

Contractors 6%

Tax & insurance 6%

Fuel & power 18%

Repairs 20%

Depreciation 50%

important if over-mechanisation is to be avoided. Power costs (see Figure 1.2) can amount to 30–40% of the fixed costs of many farms, but with labour this can be as high as 60–70%. On some farms labour and machinery costs can be a staggering 90%.

Timing
The difference between a good farmer and a poor farmer can be as little as two weeks. The farmer who is late in sowing, fertilising and applying pesticides is invariably late in harvesting. There is plenty of research which shows the importance of timing on crop response – a lack of good timing will result in reduced yields, soil damage and low or no profit.

Improved efficiency of machinery
As farm machinery costs rise, the good business manager ensures that existing equipment is working to capacity and that the whole mechanisation system, including the labour team, works as efficiently as possible.

Own equipment versus the use of contractors
As the economics of farming change, there is a need to consider all aspects of machinery use. Knowing the exact costs associated with owning and operating a machine, within the framework of good timeliness, will allow comparisons to be made with the alternatives to ownership such as contractors and machinery rings.

Effect on other aspects of the farm
Improving the output of one machine may have a serious knock-on effect to other aspects of the business. The adoption of an improved cultivation method, for example, may allow a greater area to be cultivated or, alternatively, better timeliness. The converse is the example of buying a high output combine whilst still using a small grain reception pit and associated handling facilities.

Increased area
Planning for future possibilities is a key to good management. It is a disaster to invest in a combine harvester of a given size, only to discover that it is too small when the opportunity to take on extra land arises a year later.

Decreased labour costs
Labour costs, particularly the background ones such as insurance, social security and housing costs continue to increase. No manager likes to reduce their workforce but mechanisation planning allows the farmer to take advantage of the opportunity which may arise when an employee retires.

Awareness of soil damage

The greatest asset a farmer has is the soil. Farmers must plan for that one year in ten when access to fields may be difficult due to bad weather. Large diameter tyres on trailers, avoiding 'recreational tillage' and correct capacity equipment to ensure timeliness should all be considered in mechanisation planning.

Damage to the crop

Harvest damage can be greatly reduced by using modern design and correct operational techniques (e.g. a correctly spaced lifting web operating at the correct speed can considerably reduce tuber damage on a potato harvester).

Matching equipment

The most expensive machine on a farm is likely to be a combine or forage harvester. Any delay in its output can lose money so good logistical support is vital, i.e. trailers and their drivers must be well organised.

Fuel efficiency

Inappropriate matching of equipment will result in a waste of fuel. A little understanding of engine power/torque and soil characteristics will allow the farmer to match equipment with the tractor.

Decreased variable costs

The variable costs such as fuel, labour, repairs and maintenance need to be considered and planned for. Good operator training in routine maintenance and servicing can be planned.

Records will monitor use and pinpoint expenditure

Records will, for example, allow the manager of machinery to know the true costs of ownership, to flag up service dates and to highlight excessive expenditure. They must be straightforward to complete if they are to be used and accepted by staff.

Depreciation

The cost of depreciation can be one of the largest components of the cost of owning machinery; the development of a planned, flexible replacement policy is a must. Farmers can do much to reduce the physical rate of depreciation by looking after their equipment and obtaining a good resale value.

Taxation

Forward planning to replace equipment, as and when necessary, is very important; replacing equipment purely on taxation grounds is absolutely ridiculous. Advice on taxation is only one of the many criteria which should be considered when drawing up a forward plan.

Health and safety

As health and safety legislation continues to be developed and implemented on the farm, so the farmer needs to plan ahead. Legislation is rarely implemented without notice; forward planning to take account of recently introduced legislation is part of good management (e.g. purchasing a new sprayer fitted with a low-level induction bowl will reduce operator contamination).

Cashflow planning

All businesses must be aware of the costs of production and their projected income; farming is no exception. Mechanisation planning is one of the aspects of good business management that will allow the farmer to make informed decisions about the business.

Future developments

As technology develops to improve farming, so the farm business must be profitable if the farmer is to continue to invest in improved techniques of production. To remain competitive in a world market the farmer must consider technology as one of the aids to good farming.

Web sites

http://www.adas.co.uk
http://www.aea-online.org.uk/
http://www.farmgate.com/
http://www.farmline.com
http://www.fwi.co.uk/live/
http://www.iagre.demon.co.uk/
http://www.landsmans.co.uk/
http://www.maff.gov.uk/
http://www.members.rbnet.co.uk/
http://www.rase.org.uk/
http://www.Profi.com
http://www.sri.bbsrc.ac.uk/

CHAPTER 2

SYSTEMS ANALYSIS I:
THE COMPONENTS

Farmers are faced with many changes in the farming scene. A reduction in the price of farm produce and uncertainty about future EU legislation have resulted in a greater need for attention to detail when examining the farm business.

This chapter looks at systems analysis in mechanisation management, showing how farmers can identify problem areas and improve the efficiency of their mechanisation system. The analysis of a harvesting system is used as an example.

THE IMPORTANCE OF SYSTEMS ANALYSIS IN MECHANISATION MANAGEMENT

Agriculture, like many other industries, faces economic problems. Over-production has resulted in the introduction of government and EU legislation to control prices and production. The value of most major crops and livestock enterprises is falling, the costs of production are rising and the result is a decline in profits. It is against this background that farmers must consider the fixed costs of production very carefully, and, in particular, machinery costs.

Design Principles

Systems analysis in mechanisation management follows the principles of disassembly of the harvesting system and then the assembly or synthesis to produce a better harvesting system. Attention to detail when analysing the harvesting system will enable the farmer to question and then develop individual areas within the system, thus allowing improvements to be made.

Figure 2.1 shows the components of a typical cereal harvesting system. The system encompasses many aspects of harvesting, from biological characteristics of the crop, through the physical constraints of the machine, to the attitude and motivation of the operator.

Combine harvester capacity is based upon a number of inter-related factors including:
- physical constraints
- operational constraints
- environmental constraints
- biological constraints

Traditionally, speed, width and efficiency have been the major areas to

7

consider but research has shown that drum width, straw walker area and management are of at least equal importance. The introduction of the Case I-H 'Axial-flow' and subsequent 'rotary' type combines with their increased outputs has resulted in a greater interest in combine selection.

Machine design has changed dramatically in recent years. Recent trends in design have led to a better working environment for the operator, both in driving the machine and in the machine's servicing. Driver comfort has improved considerably with the advent of air-conditioned 'quiet-cabs' and tinted glass. The use of electronics to monitor shaft speed and grain loss has greatly improved output, particularly as the driver can be quite isolated in a 'quiet-cab'. The better environment means less stress for the operator who remains alert for longer, creating an improved attitude to the length of the working day.

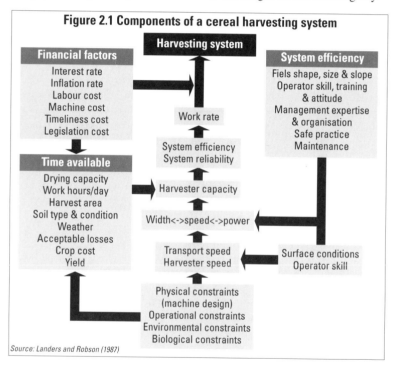

Figure 2.1 Components of a cereal harvesting system

Source: Landers and Robson (1987)

Servicing characteristics have also improved, enabling quicker, easier servicing, and an improved attitude from some operators to this task.

Harvester speed is a particularly important variable factor. Harvester speed is affected by surface conditions (e.g. seedbed preparation such as level ploughing), and the operator's attitude within the bounds of safety. The advent of the hillside combine has increased both operator safety and increased the harvest area for many farmers.

Transport speed will obviously affect the output of the combine and delays in unloading combines must be avoided. Road surface conditions and access to fields must be good and it is important to have good, safe drivers on tractors with efficient brakes.

To obtain the optimum performance from the harvester, the operator must use the power of the engine. Good maintenance will ensure that air cleaners, radiators and lubricants are checked according to the manufacturer's instruction book. Power requirements can be quite high, particularly, for example, when straw choppers are being used.

The biological characteristics of the crop being harvested will affect harvester output although of course characteristics will vary from region to region, depending upon the season and the variety being harvested.

The time available for harvesting will be dependent upon a number of factors. The weather, along with soil type and condition, will affect the harvest. The start of harvest will depend upon the acceptable level of grain moisture content, the drying capacity and the area to harvest. Forecasting by the Meteorological Office, based upon historical and present data has improved greatly in recent years and the Internet can provide the latest forecasts and therefore help planning.

As the season progresses, so the effect of grain loss increases due to over-ripening and laid cereals. The farmer's attitude to risk, based upon crop losses, crop cost and returns versus combine output and dryer capacity must be considered, as should the soil type, its effect on planting dates, crop growth and soil compaction. The amount of time available for work within the day is also a factor. The workday can be lengthened by the use of relief drivers at meal times, and an improved attitude from staff to working overtime. The length of the workday is also affected by vehicle lights and grain drying facilities.

Labour and machinery planning must be considered alongside factors such as maximising profits, improving farm standards and timeliness. The quality and quantity of information for planning labour and machinery in the past was limited and left a lot to be desired. Modern methods of planning, using computers, can increase the speed of what traditionally was a very lengthy process.

Financial factors will have an important part to play in the selection of equipment for the harvesting system. The cost of ownership, based upon capital cost, interest rates and depreciation and running costs, must be compared to the alternatives such as hiring, leasing and contracting.

Hiring has developed greatly in the last decade. The cost of hiring is quite similar to that of leasing a combine, with the added advantages of a 24-hour breakdown service and a new, modern machine every year.

Good planning of the harvest with a contractor will ensure a good service, providing he is part of the system and not just called upon in emergencies. These alternatives are discussed in greater detail in Chapter Ten.

The timeliness cost, that is, the cost of delaying harvest until the grain is dry (and subsequent shedding losses incurred) must be considered alongside the costs of drying the grain. The other important timeliness costs are those following the harvest: the cost of delayed planting and the subsequent reduction in yield. Good records will enable the farmer to work out the probable costs.

The system efficiency and reliability will affect the output of the harvester. The field's physical characteristics (size, slope, shape and soil type) will affect output. Field size coupled with grain trailer size and number will affect the transporting of grain. Field slope will affect output due to wheel grip, driver comfort/safety/risk and grain loss due to the machine's design and its operation on slopes. The use of hillside combines and more recently, optional up-hill kits, can help reduce grain loss and improve output.

Operator skill and attitude will affect workrate. A well-trained operator will make better use of expensive equipment, resulting in optimum output under the prevailing conditions. Manufacturers and organisations such as ATB Landbase offer operator training.

Good management of the harvesting team is essential; attention to detail and motivation of the staff will help obtain the best performance. A conscientious labour manager can have a great influence on the harvesting team.

All equipment is potentially dangerous. Employee education in safe practice and what can be done to prevent accidents is vital. The Health and Safety Executive publish many leaflets to keep operators informed about safety.

The Decision-Making Process

Mechanisation management is concerned with both long-term and short-term developments on the farm. Figure 2.2 shows a decision procedures flow chart. The use of a flow chart will help the farmer with the decision-making procedure as it itemises various components of the harvesting system. The farmer or adviser has to identify the problem(s) limiting a system's output or performance, whether physical or financial, based upon previous experience.

Farmers should compare a system's performance against standard data such as capacities, output, labour use and timing. Having obtained the necessary information via reports, journals, the Internet or advisers, farmers can evaluate new ideas and then revise the system accordingly. Farmers must be aware of system constraints such as financial limitations, for example available capital, interest rates and depreciation and physical limitations such as labour requirements, field size and shape. Once new ideas have been implemented, there should be constant monitoring of the system to ensure that new performance levels are being maintained.

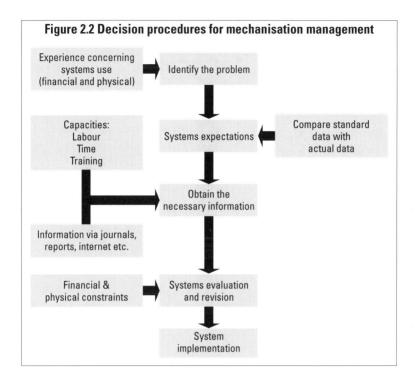

Figure 2.2 Decision procedures for mechanisation management

Short-term goals, such as improving margins, must be seen within the context of the long-term objectives of the farm business. Maximising profit within the context of land, labour and capital constraints is the most important objective of any business.

Further reading

Barnard, C.S. and Nix, J.S. *Farm Planning and Control.* Cambridge University 2nd edition 1979.

Landers, A.J. and Robson, T.F. *Systems Analysis in Mechanization Management.* In: Videotex, information and communication in European agriculture: Proc. 15th Symposium, European Assoc. of Agric. Economists, Kiel, Germany, February 1987. (G.Schiefer ed.) pp 205–224.
Kiel: Wissenschaftsverlag Vaulk.

CHAPTER 3

LABOUR AND MACHINERY PLANNING

WHY IS IT NECESSARY?

Changes in cropping policy
A farmer may decide to change cropping policy as a new market has been developed. The farmer needs to know what effect this will have on existing labour and machinery and if his existing equipment will be able to cultivate/sow/harvest etc within a given time. The farmer also needs to know the knock-on effect of introducing a new crop.

New enterprises
Herb production for ethnic groups in the Midlands may, for example, be the latest venture promoted at management conferences. Growers need to know what effect the introduction of such an enterprise would have on existing labour and machinery.

Changes in techniques of production
A farmer has a good relationship with a major supermarket which purchases potatoes. They no longer wish to purchase potatoes treated with a sprout suppressant and would like the farmer to erect a controlled atmosphere store and pack potatoes in a new packing system. The farmer needs to know the effect on the farm of an investment in controlled atmosphere storage and improved grading lines.

New or better machines
The introduction of large single-pass cultivator drills can reduce the number of cultivation passes and improve depth of drilling and timeliness. The farmer needs to know the effect on output and timeliness of the new machines, bearing in mind the conditions on the farm.

Changes in labour
The farmer needs to know the effect on the output of the farm team if Fred, the trusty retainer, retires after 50 years of excellent service. Similarly, what will Johnny will do when he returns to the farm after three years of agricultural education?

Financial constraints
A change in circumstances is brought about when the partnership between two farming siblings is dissolved. The farmer needs to know if the remaining share of the machinery inventory will cope with the land available or if it will have excess capacity.

If target dates are regularly missed or excessive overtime is necessary to meet them

A classic sign that equipment is under-sized or that management needs to be improved; the farmer needs to know how much larger the equipment should be in order to avoid over-mechanisation.

As a general aid to management

Every business should review its strategies, is it making best use of resources, can work be contracted in/out? 'What if' scenarios may be used to review the effect of changes on the farm business.

Changes in legislation

The government announces a ban on the use of an organo-phosphate insecticide, the only product which controls insects on the farm's crops. The farmer has to change enterprises and so needs to know the effect on existing labour and machinery.

Approaches

'Whole farm level'

The farming system is reviewed throughout the year. All field and livestock operations are scrutinised and planned over a 12-month period. This type of planning exercise may be carried out when a major change occurs, such as taking on a new farm or changing farming policy. There are always people who wish to farm, who are successful in obtaining land and therefore need to draw up a forward plan.

'Operation level'

A review of an existing operation is carried out to improve the performance of specific tasks (e.g. silage making, grass harvesting, etc). A very common example is the potential introduction of a 'new, improved and larger machine'.

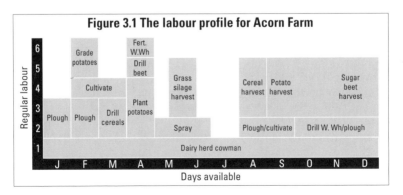

Figure 3.1 The labour profile for Acorn Farm

Figure 3.1. uses Acorn Farm to show a typical labour profile on a mixed farm. A mix of operations shows the peaks and troughs of a variable labour demand. The spring period is spent cultivating soil prior to the drilling of cereals and sugar beet and the planting of potatoes. The equally busy autumn period shows the clash between the cultivations for the winter wheat seedbed and potato and sugar beet harvests.

Planning

Collect data – questions & answers about the main operations on each crop

Establish target dates, such as drilling dates, harvest dates and time to be on/off the land. Establish the sequence of operations for each crop (e.g. rotations and seedbed establishment methods). Establish labour availability by constructing a labour profile chart (Figure 3.2). To establish the workdays available information may be obtained from farm management handbooks, Soil Survey publications or farm diaries/records on how many field workdays are available each month for the soil type and location.

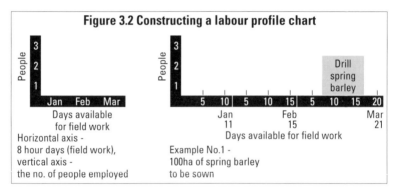

Figure 3.2 Constructing a labour profile chart

Days available for field work

Horizontal axis - 8 hour days (field work), vertical axis - the no. of people employed

Days available for field work

Example No.1 - 100ha of spring barley to be sown

Example One

A farmer wishes to sow spring barley and notes the following information:
- 100ha of spring barley to be sown.
- drilling to be carried out by a two-man team; one drives, one keeps the seed trailer filled and positioned on the headland.
- time: mid-March
- diaries indicate output of 10ha per day

To plan labour and machinery requirements accurately the farmer should build up a profile of the major operations throughout the year, e.g. following drilling there may be harrowing or rolling of the seedbed, followed by top-dressing, then pesticide applications and then harvest, baling straw, etc. Build up the profile with other activities going on throughout the farming year (see Figure 3.1).

Example Two
To construct a harvesting period on the profile, the farmer notes the following:

100ha of winter barley to harvest

team	= 1 combine driver, 1 tractor and trailer driver
when	= mid-July start
output	= 1 combine @ 1.5ha/hour output
	= 66 hours to combine 100ha
Average	= 10 hours/day
	= 6.5 days to combine 100ha

Example Three
To calculate the number of people required and the time needed to plough 58ha of oilseed rape stubble, the farmer notes the following:

plough output	= 4.8ha/day, each man
period available	= 6 days in August
4.8ha/day x 6 days	= 28.8ha/man
	= 58ha requires 2 men

The *Farm Management Handbook* states that the premium farmer's output = 2.0 man hours per ha to plough stubble.

58ha	= 116 man hours
$\frac{116}{8}$	= 14 standard man days
days available	= 6
	= 2.3 men needed

Gang workday charts
The use of a labour profile chart identifies peaks and troughs but on many farms there may be a need to know the number of people needed to form a work gang (e.g. teams for silage making, grain harvesting, potato and sugar beet harvesting). Field-scale vegetable growers need to know gang sizes/outputs for numerous field operations and packhouse jobs.

The gang workday chart is used to depict the size of gang within a given time (e.g. 80 work hours could mean one person for 80 hours or a gang of ten people for eight hours). Eight-hour days are used as the standard workday. The gang workday chart is similar to the labour profile discussed earlier: it comprises the number of days available for field work on the x-axis and the gang size on the y-axis.

A chart can be drawn up from a simple table, created from data in the

Harvest	Area (ha)	Time (approx)	Team size (ha/day)	Work rate	Gang work days (GWDs)
Winter barley	75	Mid July–mid Aug	3	10	7.5
Winter wheat	75	Mid-Aug–mid Sept	3	10	7.5
Baling/carting straw	150	Mid-July–mid Sept	3	12	12.5
Sugar beet	20	October	3	8	2.5
Field work: Ploughing	60	October	2	12	5
Drilling	60	October	1	16	3.75

Table 3.1 GANG WORKDAYS FOR OAK TREE FARM - JULY TO OCTOBER

farm management handbooks mentioned on page 17. The example shown in Table 3.1 and Figure 3.3 is based on Oak Tree Farm, a 200ha arable farm cropped with 150ha of winter barley and winter wheat and 50ha of sugar beet and shows the harvest period from mid-July to the end of October.

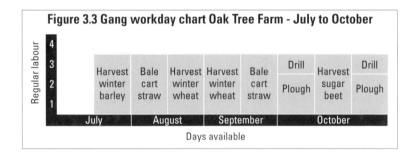

Figure 3.3 Gang workday chart Oak Tree Farm - July to October

Time available for fieldwork

Days available for fieldwork are dependent on such factors as soil type and condition, topography, climate, crop type and field operation, machine type and attitude to soil compaction and product quality.

Farmers must consider when conditions are ideal for many field operations. Spraying, for example, is now controlled by conditions and an assessment under LERAP (Local Environmental Risk Assessments for Pesticides), where drift hazards are assessed according to the wind speed, nozzles selected, etc. The weather has the greatest effect; rain prior to spraying will affect moisture on the leaf surface, rain during or after spraying will wash off the pesticides. Too high a temperature will lead to evaporation of droplets and too high a wind will lead to drift. It is little wonder that there are only a handful of officially classified 'suitable' days for spraying.

Harvesting grains or grass requires a sequence of dry days but most farmers will recall a season where harvesting continued in inclement conditions. Problems of defining workdays are also compounded by situations where the crop is 'fit' to harvest but the weather isn't suitable and vice versa.

Sources Of Information

The best source of information can be found in farm records/diaries from previous years' activities. Many research projects have been undertaken to consider the probabilities of good weather for fieldwork. Historical weather data can be readily obtained from airports for example, but an increasing number of farmers keep quite detailed records.

The Meteorological Office can provide local historical information. ADAS and MAFF are also a good source. Weather data may also be found on a number of Internet sites.

Farm management handbooks from the Agricultural Economics departments of the major British universities also contain useful data from surveys on farm business data. For example, farm investment and output in North-West England is reported by the University of Manchester while Central Southern England is reported by the University of Reading. An extremely detailed report on machinery investment in Eastern England may be found in the report on farming in the eastern counties of England, published by Cambridge University. Two very popular farm management reports are published and sold as handbooks from Wye College and the Scottish Agricultural Colleges.

A very localised and useful source may be found in the *Soils and their use* series published by the Soil Survey. Land is classified for its suitability for growing certain crops and excellent charts may be found for particular soil types.

Figure 3.4 shows how difficult it can be to obtain 60% productive hours. It is no easy matter and requires good, well-motivated people – both staff and management – to maintain a high level of productivity. Whilst not every farm has the travel requirements usually associated with a contracting business, a number of large farms and estates have a fair number of roads and tracks.

Table 3.2 gives a guide to the average days available for fieldwork in a month. Workdays will vary considerably and, for planning purposes, farmers must consider all the factors discussed in this chapter. I am reminded of the farmer I met in Norfolk, working on light sand soils, who said they worked on the land every day except Christmas Day. I also know farmers in the Thames Valley whose silage equipment has been bogged down in a field in mid-July.

Farmers must consider the weight of the equipment, its tyres and the soil condition (Figure 3.5), e.g. a low ground pressure vehicle may travel the fields most of the year. Whilst average workdays may be used for

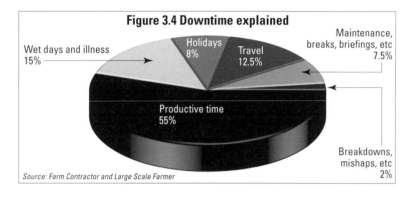

Figure 3.4 Downtime explained

Wet days and illness 15%

Holidays 8%

Travel 12.5%

Maintenance, breaks, briefings, etc 7.5%

Productive time 55%

Breakdowns, mishaps, etc 2%

Source: Farm Contractor and Large Scale Farmer

Table 3.2 AVERAGE DAYS AVAILABLE FOR FIELD WORK

Month	Light	Heavy
January	23	13
February	22	13
March	24	16
April	26	17
May	27	20
June	28	23
July	29	24
August	27	23
September	26	22
October	25	20
November	23	18
December	22	13

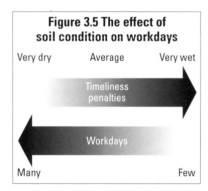

Figure 3.5 The effect of soil condition on workdays

Very dry   Average   Very wet

Timeliness penalties

Workdays

Many   Few

planning, many farmers retain an old tractor or two at the back of the barn to help out during that one year in ten when conditions are awful! Good record keeping is a must even if one year in ten may be the wettest on record.

Good planning is impossible unless there is a supply of good data. As decision-making methods become more and more sophisticated, the requirement for accurate, reliable data increases. The best data is the farmer's own records/field diaries, but, if one is planning a new enterprise or purchasing a new machine, then data from an independent, reliable source is necessary.

There are many sources of data, ranging from the departments of Agricultural Economics at the British universities who contribute to the Farm Management Survey, through surveys conducted by the Royal Agricultural Society of England, to information from land agents, accountancy companies and consultants. Three of the most popular data sources for comparison and planning purposes are:

The Farm Management Pocketbook, Wye College (Nix 1998)
Farm Machinery Costs, Agro Business Consultants (ABC 1998)
Farm Management Handbook, Scottish Agricultural Colleges (SAC 1998)

How useful is standard data from management handbooks?

An exercise was conducted examining the suitability of standard data for planning purposes. Three farms were chosen: a large arable farm, a small arable farm and a mixed farm. The farms, in different areas of the country, had differing soil types, topography, climate and husbandry practices. The three farms were chosen as the author knew the farmers were good record keepers.

South Farm
A 260ha farm situated on very heavy clay in North Wiltshire on the Thames floodplain. The cropping policy is 200ha winter wheat with a break crop of 58ha of oilseed rape.

Sherbury Park Estate
A 640ha Cotswold estate situated 200–305 metres above sea level. A mixed farm, it comprises sheep, dairy cows and arable crops. Due to its exposed situation, the crops of cereals, grass, and oilseed rape are later than surrounding farms.

Pilton Farm
A heavy clay farm situated in a low rainfall area in Bedfordshire with a simple cropping policy of 720ha of winter cereals and 200ha of oilseed rape.

Data was taken from four information sources:

Nix (1998)
One of the most popular and comprehensive sources of data, the *Farm Management Pocketbook* was first published in 1966 and is based upon results of the farm management survey from Wye College for farms in South-East England.

Walford (1979)
Walford carried out a survey of 62 arable farms in excess of 300ha in South-East England as part of a report on labour and machinery use.

This data source is applicable to two of the three farms.

Culpin (1976)
A popular, traditional source of data which was synthesised from field studies and operations by the Ministry of Agriculture's Advisory Service. It was chosen to see if farming has become more efficient.

Hunt (1983)
Professor Hunt was based at Iowa State University in the USA and provided data which allows workrates to be calculated from first principles. The efficiencies and forward speeds are based upon studies of field operations on large arable farms in mid-west USA. However, the value of this information is questionable as it may not be applicable to British farms.

Tables 3.3 and 3.4 show that substantial differences can occur between actual and standard data. Table 3.3 shows the differences in workrates and Table 3.4 the differences in labour requirements.

Table 3.3 STANDARD DATA ON MAN HOURS/HECTARE COMPARED WITH FARMERS' OWN DATA							
Operation	South Farm	Sherbury Park Estate	Pilton Farm	Nix	Walford	Culpin	Hunt
Fertiliser spreading	0.53	0.3	0.2	0.32	0.4	0.32	0.23
Spraying	0.2	0.26	0.2	0.32	0.32	0.36	0.23
Combining	2.5	2.5	1.6	2.0	2.85	2.4	3.7
Spring tine cultivation	-	0.2	0.16	0.4	0.53	0.3	0.4
Ploughing	1.67	1.1	-	1.3	2.0	1.3	1.6
Discing	0.8	0.25	-	0.6	0.9	0.6	0.4
Power harrowing	1.25	-	-	0.9	1.25	-	-
Harrowing	-	0.2	-	0.3	0.4	-	0.32
Drilling	0.57	0.5	0.5	0.4	0.6	0.5	0.5
Rolling	0.5	0.3	0.18	0.4	0.5	0.4	-
Mow & Forage Harvesting	-	6.85	-	5.8	-	1.6	2.2

Actual work rates will vary according to field shape, size, slope, efficiency and soil type. The equipment, its width, organisation, operation, and condition will affect work rates. Systems analysis is discussed in some detail in Chapter Two.

Standard, reliable data has its uses, it is certainly better than the alternative: no data! Standard data must be adjusted for local conditions, personal experience, etc. It is best suited for simple, straightforward enterprises, as found on arable farms. The complexity of operations and labour requirements on mixed farms can lead to very strange results, as shown in Table 3.4.

Reliability of data is important. For example, is it based upon the best work rate for one day or the average work rate for a season? Great care must be taken to ensure that the data is useful.

Table 3.4 COMPARISON OF STANDARD LABOUR DATA WITH LABOUR EMPLOYED					
Farm	Actual labour team	Nix	Walford	Culpin	Hunt
South Farm	4	3	4	4	5
Sherbury Farm Estate	14	18	22	21	28
Pilton Farm	7	6	7	7	8

Change in policy

The Acorn Farm example below is described in Figure 3.1 (see page 13).

The busy periods described in Figure 3.1 were worrying the farm manager. Organising labour and machinery in such a busy time, concerns about timeliness, soil damage and overtime pay resulted in a policy change. With one employee due to retire, the sugar beet equipment quite old and needing major capital injection it was decided not to replace the retired employee and to grow oilseed rape instead of sugar beet.

Figure 3.6 shows the change on the profile chart. The oilseed rape could be drilled and harvested with existing equipment used for cereals. The spring and autumn clashes have been reduced as sugar beet was eliminated. Soil damage has been reduced considerably as heavy harvesting equipment and trailers are no longer on the fields in November and December. The slack time in July is now taken up in mid-July with the harvest of oilseed rape. The only disadvantage is the disruption to the cereal harvest team for three days as one person goes drilling oilseed rape in late August.

The advantages of drawing up a labour and machinery planning chart are that the chart:

● highlights busy and slack periods (as peaks and troughs) and can be used to improve levels of mechanisation to ease peaks and troughs

● identifies labour requirements and therefore number of tractors required. Would the employment of a contractor or casual labour alleviate any problems?

● can be drawn up fairly rapidly, particularly using graph paper or cells in a computer spreadsheet

● can be used to plan future requirements such as changes in labour, machinery output, farm policy and area farmed. Would such changes enable a more even demand throughout the year?

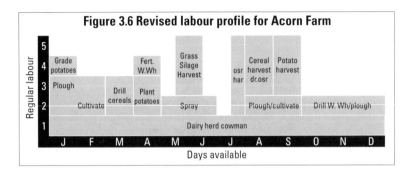

Figure 3.6 Revised labour profile for Acorn Farm

The disadvantages of drawing up a labour and machinery planning chart are:

- it is not a precise technique – only a guide due to differences in standard and actual data, seasonal variability and time available to perform field operations
- time required to construct may be a problem although this depends on the complexity of the farming business
- the value of the results obtained depends on the assumptions made

Further reading

Bailey, J. and Graham, K. How Much Machinery Do You Really Need? *Farm Management* Vol. 6 No.10 Summer 1988.

Barrett, F.M., Arable work scheduling and projection: an adviser's method. *The Agricultural Engineer* Vol.37 No. 4 1982.

Cracknell, J. Factors influencing the mechanisation of UK agriculture since 1972. *The Agricultural Engineer* Vol.49 No.3 1994.

Errington A. Developing talents and team building. *Farm Management* Vol.8 No.3 Autumn 1992.

Farm Management Handbook, SAC, Edinburgh, 1999.

Farm Machinery Costs, Agro Business Consultants, Melton Mowbray, Leics.

Gladwell,G. Labour and machinery analysis and planning. *Farm Management* Vol. 5 No.9 Spring 1985.

Hill, P. *Farm Management Pocketbook*. Wye College, University of London,1999.

MAFF/PSD, *Local Environmental Risk Assessments for Pesticides*, MAFF Publications, London.

Oskoui, K.E. The practical assessment of timeliness penalties for machinery selection purposes. *The Agricultural Engineer* Vol.38 No.4 1983.

McGechan, M.B., Saadoun, T., Glasebey, C.A., Oskoui, K.E. 1989 Estimation of combine harvesting workdays from meteorological data. *The Agricultural Engineer* Vol.44 No. 3.

McGechan, M and Cooper, G. 1994.Workdays for winter field operations. *The Agricultural Engineer* Vol.49 No.1.

Murphy, M.C. *Report on Farming in the Eastern Counties of England*. Dept. of Land Economy, University of Cambridge. 1999.

RASE & Andersons. *Model Farm Business projects – Brash and Loam Farms*. Stoneleigh, 1997.

Soil Survey. *Soils and their use series*, Silsoe: Soil Survey and Land Use Centre. 1985.

Vaughan, R. *Farm Business Data 2000*. Dept. of Agricultural and Food Economics, University of Reading. 1999.

Yule, I. Copland, T.A.and O'Callaghan, J.R.O. Machinery utilisation on arable farms. *The Agricultural Engineer* Vol. 43 No. 2 1988.

Web sites
http://www.agrilink.co.uk/
http://www.farmgate.com/
http://www.fwi.co.uk/live/
http://www.meto.govt.uk/sec3/sec3.html
http://www.silsoe.cranfield.ac.uk/sslrc/
http://www.silsoe.cranfield.ac.uk/sslrc/services/

CHAPTER 4

WORKRATE

In Chapter Three it was shown that the construction of a labour and machinery profile depended upon a number of factors, including the workrate of farm machinery. Workrate figures are used by farmers to determine output when comparing one machine against another and for mechanisation planning.

There are a number of rates quoted:

- spot workrate
- net workrate
- overall workrate
- seasonal workrate

It is very important to know which workrate is being quoted if one is to avoid an oversized or undersized machine.

Spot workrate

Spot workrate is the rate of work of a machine constantly in work, i.e. without stopping to turn around or for repairs, maintenance and adjustments. The best example is the unlikely situation of someone mowing the central reservation of the M1 motorway from London to Leeds without stopping. The rate of work would be fantastic!

Spot workrate can be used to compare two machines working in a field, such as at a ploughing or cultivation demonstration. It is useful for side-by-side comparison. It is important to note that spot workrate is often used by salesmen as it is the highest workrate and therefore the most impressive. Fortunately our fields aren't the length of the M1, so turning and stoppages must be taken into account.

$$\text{Spot rate} = \frac{\text{speed (km/hr) x width (m)}}{10} = \text{ha/hr}$$

e.g. tractor and 4m discs travelling at 8km/hr

$$\frac{8 \times 4}{10} = 3.2 \text{ ha/hr}$$

Net workrate

The net workrate includes turns but doesn't take into account other stoppages such as machine adjustments and maintenance. Net workrate usually allows about 10% of time for turning. It is of little value in planning.

Overall workrate

Overall workrate is the most popular workrate in use because it is the most realistic; it takes into account the efficiency of the system as a

whole. Field efficiency is discussed later in this chapter. It applies an efficiency factor by allowing for:
- filling/emptying
- adjustments
- routine maintenance
- stoppages such as blockages

The equation used to measure overall workrate is:

$$\frac{\text{speed (km/hr) x width (m) x efficiency (\%)}}{10}$$

e.g. tractor and 4m discs, 8km/hr at 85% field efficiency

$$\frac{8 \times 4 \times 0.85}{10} = 2.72 \text{ ha/hr}$$

Seasonal workrate
Includes major breakdowns, mealtimes, travelling from field to field, weather delays and major stoppages. Based upon the machine's output spread over the season and includes meal breaks, it is the most relevant to forward planning for work schedules etc, but detailed, whole-season records, which sometimes may be difficult to obtain, are needed.

Table 4.1 AVERAGE COMBINE HARVESTER PERFORMANCE USING A 3m WIDTH OF HEADER IN BARLEY		
Theoretical spot rate of work	1.50 ha/h	–
Effective spot rate of work	1.35 ha/h	6.8 t/h
Overall rate of work	1.04 ha/h	5.2 t/h
Seasonal rate of work	0.92 ha/h	4.6 t/h
Field efficiency	69%	

Source: SAC Publication No. 88, 1982

Table 4.1 clearly shows the variation between the different types of workrates and reiterates the need for care when selecting equipment based on workrates. It would be a harvesting disaster if a combine was purchased based upon spot workrate and not overall workrate.

CALCULATING FIELD EFFICIENCY

Actual capacity of a machine = theoretical capacity (spot workrate) x field efficiency

% efficiency $\quad = \quad \dfrac{\text{actual workrate x 100}}{\text{input (spot workrate)}}$

Typical (average) efficiency figures can be obtained from Table 4.2.

More realistic figures for a particular farm may be obtained by measurement and calculation for the situation.

For example:

4m discs operated at a speed of 8km/h complete 10ha in 4 hours.

$$\frac{10\ ha}{4\ hours} = 2.5ha/hr\ input$$

Table 4.2 FORWARD SPEED AND FIELD EFFICIENCY

	Typical forward speed km/hr	Field efficiency %
Ploughing		
Conventional	5.6	70
Reversible	5.6	85
Chisel ploughing	6.4	80
Disc harrowing	8.0	85
Medium tined cultivation	8.0	85
Spring tine harrows	9.7	85
Seed harrows	8.0	85
Cambridge rolls	4.0	85
Drilling		
Grain only	9.7	70
Combine grain and fertiliser	9.7	60
Rotary cultivation	4.0	85
Reciprocating and rotary harrows	6.4	85
Mole ploughing	3.2	74
Deep subsoiling	4.0	80
Fertiliser spreader		
Small hopper 50%	8	50
Large hopper, big bags	10	75
Crop sprayer		
Mounted	9	50
Self-propelled	10	75
Mower		
Mounted	8	80
Trailed	10	85
Forage harvester		
Trailed	7	65
Self-propelled	10	80
Combine harvester	5	75
Potato planter	8	60
Potato harvester	5	65
Sugar beet harvester	6	60

These figures should only be used as a general guide

If spot workrate = 3.2ha/hr then field efficiency = $\dfrac{2.55 \times 100}{3.2}$ = 79.7%

This calculation assumed that the set of discs was working at full width all the time, hence the spot workrate assumed that the working width was 4m. But the working width of most farm equipment (e.g. harvesters, cultivators, etc) is less than its working width. The field efficiency figure takes this into account along with turning, filling or emptying.

Theoretical And Actual Capacity

To calculate how many minutes it will take to spray a field:
Minutes/ha = 600 divided by swath width (m) divided by speed (km/h)
e.g. 18m sprayer travelling at 9.7km/h

$$\text{Minutes/ha} = \quad \dfrac{600}{18} \quad = \quad 33.3 \quad = \quad \dfrac{33.3}{9.7} \quad = \quad 3.44$$

Sprayers need to turn at the headland, refill with water and pesticides and open/fold booms so the actual capacity will be quite different to the above formula which only takes into account speed and width. An efficiency factor of 60–75% needs to be included. The efficiency will vary according to such factors as:
- boom folding – manual or hydraulic
- fill rate – small or 120mm hose with high output centrifugal pump
- rate of chemical fill – small or large containers
- people – one person or two, the second on a water bowser

Chapter Six details sprayer performance and factors affecting output.

MEASURING SPEED

The proof meter on the dashboard of a tractor often bears little relation to the actual speed of the machine. Modern tractors with radar or electronic speed indicators are more realistic than older tractors. Tyre sizes may differ from the size used to calibrate the tractor as tyre sizes will vary under load and tyre pressures may be incorrect.

Example
A tractor has travelled 120 metres forwards in one minute.

In 1 hour the distance travelled will be 120 x 60 = 7200 metres
There are 1000 metres in 1km so: 7200 divided by 1000 = 7.2km/hr

Method A
Speed (km/hr) = $\quad \dfrac{\text{distance (m)}}{\text{minute}} \times \dfrac{60 \text{ mins}}{1 \text{ hour}} \times \dfrac{1 \text{ kilometre}}{1000\text{m}}$

$$= \quad \frac{1000}{60} \quad = \quad 16.7$$

speed (km/hr) $\quad = \quad \dfrac{\text{distance (metres)}}{\text{time x 16.7}}$

Method B
Measure 100 metres in the field
Check the time taken to travel this distance

Speed (km/h) $\quad = \quad \dfrac{360}{\text{seconds to travel 100 metres}}$

e.g.
Speed (km/h) $\quad = \quad \dfrac{360}{80} \quad = \quad 4.5$

Example
The farmer at Acorn Farm decides to increase the potato area to 160ha so he needs to know the minimum size of planter to buy.

160ha of potatoes need to be planted in a period of seven working days. Each day has 12 hours of field time.

Forward speed = 6.4km/hr
Row width = 1.0m
Efficiency = 70%

Minimum overall workrate needed
Capacity $\quad = \quad \dfrac{160\text{ha}}{7 \text{ days x } 12\text{hrs/day}} \quad = \quad 1.9 \text{ ha/hr}$

Using formula for overall workrate (actual capacity)
Capacity $\quad = \quad \dfrac{\text{speed x width x efficiency}}{10}$

So: Width $\quad = \quad \dfrac{\text{capacity x 10}}{\text{speed x efficiency}}$

$$\dfrac{1.9 \text{ x } 10}{6.4 \text{ x } 0.7} \quad = \quad 4.24$$

4.24m @ 1m/row = select a 5-row planter with 1m row spacing

The effect of field shape and size on output of machines
Figure 2.1 (page 8), shows that there are numerous, interwoven factors which affect machine output. Field shape, size, terrain, obstructions and

the size of the equipment play an important part. A 24m sprayer operating in a small 2ha field which has a steep slope on wet clay soils and contains two small ponds and an electricity pylon will have a different output compared to spraying a flat, 15ha field in the Fens without any obstructions.

Two extremely good papers by Sturrock, Cathie and Payne (1977) and Sturrock and Cathie (1980) considered economies of scale and the impact of increasing field size on arable farms. Their research showed the effect of field shape and size upon the workrate of implements.

Field size and implement width
When cultivating a field the tractor and implement has to turn around at the headland. If the hedge or stonewall is removed the number of turns is reduced. The larger the field the more time can be spent in effective work (Table 4.3).

Table 4.3 PROPORTION OF A TRACTOR DRIVER'S TIME SPENT ON EFFECTIVE WORK (Implement 3m wide moving at 6km/hr)							
Field Size (hectares)	2	4	8	10	20	40	80
Effective work	37	47	57	59	65	71	74
Turns	20	19	15	14	12	8	7
Headlands	4	3	3	3	2	2	1
Changing fields	22	14	8	7	4	2	1
Contingencies	17	17	17	17	17	17	17
	100	100	100	100	100	100	100
Hectares per day	4.64	5.96	7.11	7.42	8.27	8.94	9.39
Ha/day (2 ha = 100)	100	128	153	160	178	193	202

Source: Sturrock

Wide combine harvesters are unable to realise their full potential in small fields. Table 4.4 shows that the output of a 4.5m combine harvester is much greater than a 3.3m combine when harvesting a 5ha field. Forward speed was 4.3km/h in a light/medium crop.

However, the wider table or header would need removing for transport on the highway and the time (and frustration) needs to be taken into consideration.

Increasing field size beyond 10ha to 20ha can result in an increased output of 11% for a 3m implement. The effect of increasing field size has less effect over 20ha, e.g. increasing from 40ha to 80ha only increases

Table 4.4 COMBINE HARVESTER OUTPUT

Effective width	3.3m	4.5m
Cutting time (mins)*	170	125
Turns	20	15
Miscellaneous delays	30	30
Total (mins)	220	170

* excluding headlands

Source: SAC Publication No.88

Table 4.5 INCREASE IN TRACTOR DRIVER'S WORKRATE DUE TO LARGER FIELDS AND WIDER IMPLEMENTS

Implement width	Index of workrate					Increase in workrate		
Field size (ha) (acres)	4 (10)	10 (25)	20 (50)	40 (100)	80 (200)	10–20 (25-50)	20–40 (50-100)	40–80 (100-200)
3m	80	100	111	120	126	+11%	+9%	+6%
5m	76	100	115	126	134	+15%	+11%	+8%
12m	66	100	124	142	156	+24%	+18%	+14%

output by 6%. The effect of field size on output alters with implement width, e.g. the output of a 12m implement increases by 24% when fields increase from 10ha to 20ha, yet only increases by 14% when field size goes from 40ha to 80ha – a diminishing return, but still worth considering! Table 4.5 shows the increase in the tractor driver's workrate when working in larger fields and with wider implements.

Field shape
Table 4.6, in conjunction with Figure 4.1, shows the effect of field shape on the output of a cultivator. The effect of obstructions will slow down an implement considerably. While a base level of 100 is given to a square field, note how a rectangular field where the length is four times the width results in an increased output of 7% and how a field with obstacles results in a reduction of 9%.

Work pattern within a field
The work pattern within a field also has an effect on output. The use of reversible ploughs has improved ploughing output compared with conventional ploughing in lands. Ploughing 'round and round' also improves output. But reversible ploughs are more expensive, heavier and, if a tractor can pull a three-furrow reversible, it may be able to pull a

Table 4.6 EFFECT OF FIELD SHAPE ON TIME TO CULTIVATE 10ha		
Field shape	**Minutes per ha**	**Index**
1. Square	56.6	100
2. Rectangle (2:1)	54.0	95
3. Rectangle (4:1)	52.4	93
4. Standard shape*	59.5	105
5. Re-entrant side	59.1	104
6. Building plots	60.5	107
7. Obstacles in field	62.0	109

* Two sides not parallel; adopted as standard in farm models

four-furrow conventional, thus matching output! Reversible ploughs are a must for preparing certain seedbeds such as sugar beet.

The best system to choose is one which allows the operator to continue in a pattern neither stopping nor reversing nor damaging soil or implement. Too often you see misinformed operators standing on the tractor independent 'turning brake', skewing the tractor and plough through 180 degrees, relying on the weight of the plough to maintain momentum as it rotates at speed. Smeared soil is the result.

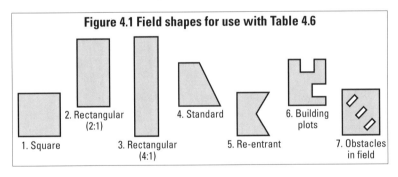

Figure 4.1 Field shapes for use with Table 4.6

1. Square
2. Rectangular (2:1)
3. Rectangular (4:1)
4. Standard
5. Re-entrant
6. Building plots
7. Obstacles in field

Harvesting work patterns also effect output. Combine harvesting in 'lands' allows trailers to pull alongside.

WORKRATE: A COMPUTERISED PLANNING PROGRAMME

Workrate computer programme can be used to see the 'what if' effects of changes in the operating parameters of farm machinery. For example:

- what is the improvement in the output of a fertiliser spreader by changing to big bags?
- what is the effect of purchasing a sprayer with a wider boom?
- what is the effect on output of moving from 18m to 24m tramlines?

Appendix A contains details of the programme for use in a computer spreadsheet. Details are given on how to enter the programme into a basic spreadsheet such as Excel, Quatro-pro or similar.

Example

The farmer at Acorn Farm wishes to calculate the efficiency of his crop sprayer system. The following information is available:

tank capacity: 1500 litres
boom width: 12.0 metres
application rate: 200 litres/hectare
speed when spraying: 9.7km/hour
speed travelling on road: 12.0km/hour (average)
average distance to water supply: 1.0 kilometre
water source: mains supply
water filling rate 1.0litres/second

Background formulae:

$$\text{Filling time} = \frac{\text{tank capacity (litres)}}{\text{filling rate (litres/min)}}$$

$$\text{Travel time} = \frac{\text{distance to and from supply (km)}}{\text{travel speed (km/h)}}$$

$$\text{Theoretical workrate} = \frac{\text{width (m) x speed (m/hr)}}{10}$$

$$\text{ha/load} = \frac{\text{tank capacity (litres)}}{\text{application rate (litres/ha)}}$$

$$\text{Time to empty tank} = \frac{\text{ha/load x 6000}}{\text{spot workrate x field efficiency (\%)}}$$

$$\text{Cycle time} = \text{time filling + time travelling + time emptying}$$

$$\text{Overall rate of work} = \frac{\text{ha/load x 60}}{\text{cycle time}}$$

$$\text{Efficiency} = \frac{\text{overall rate of work}}{\text{theoretical workrate}} \times \frac{100\%}{1}$$

Applying the observations made at Acorn Farm

$$\text{Filling time} = \frac{\text{tank capacity}}{\text{filling rate}} = \frac{1500}{1.0 \times 60} = 25 \text{ min}$$

Travel time $\quad=\quad \dfrac{\text{distance to and from supply}}{\text{travel speed}}$

$$= \quad \dfrac{2 \times 1000}{\frac{12}{60} \times 1000} \quad = \quad \dfrac{2 \times 60}{12} \quad = \quad 10 \text{ min}$$

Theoretical workrate $= \quad \dfrac{\text{width (m)} \times \text{speed (m/hr)}}{10}$

$$= \quad \dfrac{12 \times 9.7}{10} \quad = \quad 11.64\text{ha/hr}$$

Time to empty tank $= \quad \dfrac{\text{ha/load} \times 6000}{\text{spot workrate} \times \text{field efficiency (\%)}}$

$$= \quad \dfrac{7.5 \times 6000}{11.64 \times 75} \quad = \quad 51.54 \text{ min}$$

Cycle time $\quad=\quad 25 + 10 + 51.54 \quad = \quad 86.54 \text{ min}$

Overall rate of work $= \quad \dfrac{\text{ha/load} \times 60}{\text{cycle time}}$

$$= \quad \dfrac{7.5 \times 60}{86.54} \quad = \quad 5.19\text{ha/hr}$$

Efficiency $\quad=\quad \dfrac{\text{overall rate of work}}{\text{theoretical workrate}} \quad \times \quad \dfrac{100\%}{1}$

$$= \quad \dfrac{5.19}{11.64} \quad \times \quad \dfrac{100}{1} \quad = \quad 44.67\%$$

Using Workrate to perform the above calculations it would take 30 seconds to enter the information and obtain the results.

Example

The farmer at Acorn Farm wishes to improve sprayer output and efficiency and considers the following:

    increasing tank size to 3000 litres
    increasing boom width to 18m
    increasing boom width to 24m
    using a mixer wagon on the headland
    using a mixer wagon and 24m boom

Figure 4.2 shows the computer programme Workrate with the input and output of the above changes.

Figure 4.2 Workrate

Input data	Model A	Model B	Model C	Model D	Model E	Model F
Implement width (m)	12	12	18	24	12	24
Capacity (kg or litres)	1500	3000	1500	1500	1500	3000
Forward speed (km/h)	9.7	9.7	9.7	9.7	9.7	9.7
Application rate (kg/ha or l/ha)	200	200	200	200	200	200
Transport time (mins)	5	5	5	5	1	1
Filling time (mins)	25	25	25	25	10	10
Field efficiency (%)	75	75	75	75	80	80
Output						
Area covered per load (ha)	7.5	15	7.5	7.5	7.5	15
Filling rate (kg/min or l/min)	60	120	60	60	150	300
Spot work rate (ha/h)	11.64	11.64	17.46	23.28	11.64	23.28
Total time per load (min)	86.5	138.1	69.4	60.8	60.3	60.3
Overall work rate (ha/h)	5.2	6.5	6.5	7.4	7.5	14.9
Overall efficiency (%)	44.7	56.0	37.2	31.8	64.1	64.1
Components						
Application time per load (min)	51.5	103.1	34.4	25.8	48.3	48.3
Application time (%)	59.6	74.7	49.5	42.4	80.1	80.1
Filling time per load (min)	25.0	25.0	25.0	25.0	10.0	10.0
Filling time (%)	28.9	18.1	36.0	41.1	16.6	16.6
Transport time per load (min)	10.0	10.0	10.0	10.0	2.0	2.0
Transport time (%)	11.6	7.2	14.4	16.5	3.3	3.3

Increasing tank size to 3000 litres (Model B)
The output has changed very little (1.3ha/hr and an overall efficiency gain of 11.3%). The cost of adding a front-mounted tank to the tractor may well be regarded as a relatively inexpensive addition to gain improved output.

Increasing boom width to 18m (Model C)
The increase in workrate is the same as in model B but the cost of increasing the boom by 50% is certainly much higher than purchasing a front nurse tank. A 10% reduction in application time is achieved.

Increasing boom width to 24m (Model D)
This creates a tremendous increase in spot workrate, approximately double model A, and an improvement in overall efficiency of 13%. Consideration must be given to terrain and other tramline equipment

(e.g. drill and fertiliser spreader), before changing to such a wide machine.

Using a mixer wagon on the headland
Model E shows a tremendous increase in overall efficiency of 20%. The introduction of a second person, pulling a water bowser or mixer wagon along the headland helps output. The extra output, due to the considerable reduction in transport time and no boom folding, has to be weighed against the cost of a second person. Mixer tanks can be placed on a farm trailer or a purpose-made system can be purchased.

Using a mixer wagon and 24m boom
The combination of wide booms and a mixer tank results in a 300% increase in workrate. This is an ideal system where large areas of relatively flat land need spraying.

Reducing application rate
This model isn't shown in Figure 4.2, but, if conditions allow, reducing rates by 50% can improve outputs. Applying pesticides at 100l/ha in Model A improves overall workrate to 6.5ha/hr, the same improvement as was found in Model B, (doubling tank size) or Model C, (increasing boom width by 50%). On Model F, workrate increased to 16.6ha/hr.

Further reading
How to choose and use combines. Publ. No.88. SAC, Edinburgh.
Sturrock, F.G., Cathie, J., and Payne,T.A. 1977. *Economies of scale in Farm Mechanisation.* Occasional Paper No.22. Dept. of Land Economy, Cambridge University.
Sturrock, F.G. and Cathie, J. 1980 *Farm modernisation and the countryside.* Occasional Paper No.12. Dept. of Land Economy, Cambridge University.

CHAPTER 5

SYSTEMS ANALYSIS II:
ANALYSING THE SYSTEM

The 6 Ps of field machinery management are:
Proper Prior Planning Prevents Poor Performance.

ANALYSING A GRAIN HARVESTING SYSTEM

To ensure the best performance of any future harvesting system, the farmer needs to consider the faults, problems and pitfalls of the previous season. Autumn or winter is an ideal time to carry out this management exercise. There is time to plan and implement changes – and some of the points which determine crop harvesting efficiency can be affected by work done before the next crop goes into the ground.

Where were the problems? Was the combine too slow? Did the driver fail to make the most of its capacity? Was it held up because the grain trailers could not keep up? Did hold-ups at the grain store bring operations to a grinding halt? Where can the job be made easier, faster, more cost-effective?

In the context of combinable crops, the overall system can be broken down into a number of components, the most important of which are: the crop, the combine, the dryer, transport, labour and cost. Some factors affect all these elements, others are specific. For example, the type of crop grown influences decisions made across the system, while the skill of a combine operator is very specific.

At the heart of the system is the combine harvester. Its key task is to get the crop off the ground in as good a condition as possible. The efficiency of the combine harvesting operation is affected by the machine itself, the operator, the condition of the crop and the state of the field. Mechanically, combines have improved in leaps and bounds in recent years. Their capacity, reliability and efficiency have all been significantly improved and this is something which must be taken into consideration when deciding whether to soldier on with an old machine, or turn to a new (or newer) model.

Combine throughput needs to be matched to the overall area to be cut. The mechanical advances have been so great that larger farms may be able to reduce the number of combines used and smaller farms can turn to machines which are considerably cheaper to run. There are numerous examples of farmers with 300-400ha who once ran three combines and now find that one, or at most two, are adequate for their needs.

Not only have combines improved mechanically, there have also been great advances in operator comfort and design, also creating greater throughput. The modern cab reduces stress, electronics allow the machine to be driven far nearer its optimum limit and simpler maintenance requirements mean operators are far more willing to keep machines well serviced. This means the running of the combine is more cost-effective and throughput is maintained.

The skill of the operator is crucial to the system. Time and money spent in winter on training courses, such as those run by the ATB Landbase, local agricultural colleges and various manufacturers on operation and maintenance, will be well repaid. The skilled operator will make the most of the combine's potential (Table 5.1) and with training and motivation he will also look after the machine and minimise the risk of those frustrating mid-harvest breakdowns. It is not just the combine operator who needs training. The harvesting operation is full of potential hazards so attention to training in safe practices is essential for the sake of the farmer and his staff – both full- and part-time – and the efficiency of the harvest.

Table 5.1 COMBINE HARVESTER SETTINGS TO ENSURE MAXIMUM THROUGHPUT

Reel as slow as possible, consistent with good intake
    Too slow: crop is pushed away from the table
    Too fast: grains are shed onto the ground ahead of the table

Drum as fast as possible, consistent with good threshing
    Too slow: reduced throughput
    Too fast: chipped grain in the grain tank

Concave as wide as possible, consistent with good threshing
    Too close: reduced throughput and damaged grains in tank
    Too wide: unthreshed heads in the straw

Sieves as wide as possible, consistent with a good sample
    Too narrow: reduced throughput and unthreshed heads behind the combine
    Too wide: unthreshed heads and short straw in the tank

Fan speed as fast as possible consistent with a good sample
    Too slow: small, light grains and weed seeds in the grain tank
    Too fast: good grains found behind the combine

Work done long before the crop goes into the field can lead to greater combine efficiency. Even the best combine driver cannot work efficiently on rough fields. The strain on operator, machine and transport from poorly prepared seedbeds will add up to lower output and greater risk of breakdown. So be sure to get land in good condition before drilling. Attention should also be paid to roads and tracks. Make use of spare labour in the winter months by improving farm routes; it will be repaid with faster transport times at harvest.

There are biological constraints to harvest efficiency which are largely outside the farmer's control. The weather at harvest and the geographical location of the farm will have a bearing on the number of days available for harvesting. You cannot influence them but you must take them into consideration in assessing your harvest system.

If you are in a traditionally dry area you can consider a cheaper combine with lower throughput, because there will be more days available for harvesting. Drying capacity can also be less than that which a farmer in a traditionally wet area may have to consider. In some instances harvesting capacity can be improved at very little cost. For instance, a few pounds spent on spotlights to keep the combine going for an extra couple of hours each evening can dramatically affect efficiency.

Cropping pattern also influences the peak demand which the harvesting system has to accommodate. The farmer who grows nothing but winter wheat can face a very sharp peak demand for combine capacity. This can be reduced significantly by modifying the cropping mix. The farmer whose cropping area includes early ripening oilseed rape and winter barley and late ripening field beans and spring wheat will find that his harvesting period is spread out and peak demand for workforce and machinery is cut down accordingly. Even within crop types there are varieties which ripen early and those which mature later in the season, so easing harvest pressure still further.

Combine Performance (Figure 5.1)

Ensuring the combine and its operator can work at their optimum rate is only the first part of the analysis. There is no point in having a combine going flat out if the set-up for transport from the field, crop reception, drying and storage is inadequate. As cereal yields have increased by two per cent per year for the last decade or more, many grain growers find their existing facilities under pressure. An extra trailer, or larger models, can make a big difference to the efficiency of getting grain from field to store.

The distance from the outlying fields to the grain reception pit will dictate the transport needs of the system; past experience of combines standing idle with full grain tanks is a sure indication that another trailer is needed. It may be purchased, leased or just borrowed in a sharing scheme with one or more neighbours. Researchers at SAC conducted a large survey of combine outputs. They found that:

Where one tractor and two trailers are available, the grain tank size is important.

Working at 8t/hr, a medium-size combine harvester can fill its grain tank (3.2cu.m) in about 17 minutes, whilst the rival model grain tank (5.5cu.m) allows the trailer to be away for up to 29 minutes.

Clearly a bigger grain tank is likely to be beneficial.

Where one tractor and trailer is available, the trailer size is more important.

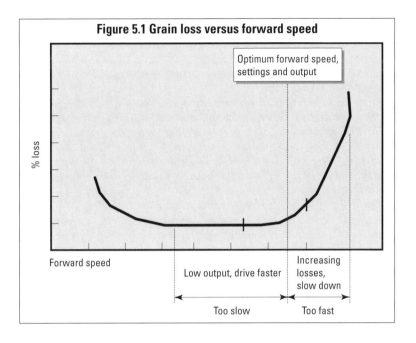

Figure 5.1 Grain loss versus forward speed

Maximum available journey time (harvesting at 10t/hr)

	2x5t.trailers	2x10t.trailers
	minutes	minutes
3t tank	48	78
4t tank	54	84

Clearly a bigger trailer is likely to be beneficial.

Deciding on drying capacity is more difficult. It demands a balance between what can be achieved in the field in most seasons and this is affected by the area's climate. Just as a balance must be struck between the loss of over-ripe grain and excessive harvesting capacity, so drying capacity must be balanced with the speed of the system.

Financial factors will naturally play a crucial role in modifying the harvest system. After all, the aim of systems analysis is to improve cost effectiveness. The two costly elements are labour and machinery, which need to be set in the context of the farm's overall financial targets and plans. If harvest has been delayed through lack of manpower there is no point in employing another man full-time for the whole year. Farmers have found that spending a couple of hundred pounds a week on relief workers – or considerably less on student labour – has met their needs without the cost of employing and housing another full-time member of staff.

Similarly, there is no need to commit oneself to buying extra equipment just for harvest. Short-term hire of combines, tractors and other costly pieces of equipment is growing in popularity. Hiring offers big advantages to hard-stretched cashflows. Similarly, there are advantages in leasing equipment rather than outright purchase for some farmers, who will find the tax benefits of two-year leases on combines attractive compared to available capital allowances.

Some farmers will need to question whether they need to be involved in the ownership, or leasing of a combine at all. A contractor may be the answer to their problems – for some smaller grain producers it certainly has been. The farmer gets a professional with all the necessary skills and equipment for zero capital outlay. Of course, calling on the contractor, like the fire brigade – only when things go wrong – is not likely to prove satisfactory to either party.

The amount of data available for analysing the financial and physical limits on a system is limited on most farms; good record keeping helps! However, the increase in computer use has meant that many farmers now have the ability to analyse their labour and machinery costs and usage carefully. Such analysis will identify where money is being wasted. The more sophisticated programmes will allow the effects of a number of alterations to the system to be considered. Alternatively, some specialist farm accountants offer such analysis at competitive rates.

Once all the factors have been analysed then the decision-making process – and the building of a better system – can begin. Some factors can be easily and quickly improved. Others will take longer to implement. The important thing is not to put off decisions. A plan to cover the next two or three harvests should be prepared, bearing in mind that flexibility is a key word in modern management.

SPRAYER OPERATION MANAGEMENT

Management of the spraying system is becoming much more important as a result of the increasing financial constraints imposed upon farming, the need for better timeliness, operator safety and stricter legislation regarding the environment.

The logistics of the spraying operation, that is, providing a good back-up service to enable the sprayer to remain in work for the maximum amount of time, is of great importance. The sprayer should be regarded as part of a spraying system rather than just a machine. Figure 5.2 shows the components of a typical spraying system. Farmers need to review the spraying operation to ensure that maximum benefit is made from the use of modern agricultural chemicals and modern sprayers. Farmers should pinpoint any aspect of the system which is slowing down crop spraying and the following should be considered:

Boom
The effect of changing boom width is only noticeable after other parts of the system have been considered, as boom width only affects the output

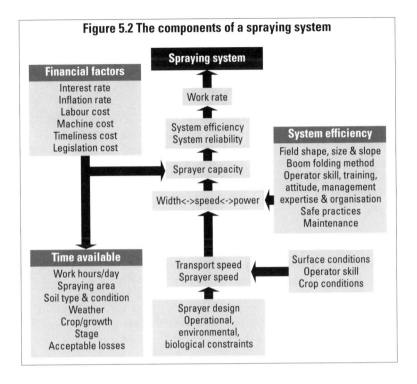

Figure 5.2 The components of a spraying system

whilst spraying. Sprayer output will increase as boom width increases due to fewer passes over the field. This leads to a reduction in the number of headland turns required, and a reduction in the time taken to empty the sprayer tank. Another advantage of using wider booms is that the reduction in wheel marks across the field means less crop damage. This is particularly important where tramlines aren't used.

Tramline width should be considered alongside boom width. The sprayer and fertiliser spreader should match the selected tramline width, normally an odd multiple of the drill width.

Wide booms without tramlines can result in problems of bout matching. Methods to aid bout matching include foam blob markers, paint sprays, string and various electronic devices.

Wider booms increase the weight of the sprayer, causing soil compaction, particularly when the soil is wet and heavy, although the use of wide tyres can reduce this greatly. Wide booms also require very good suspension – the increase in width increases the amount of boom roll and yaw. An increase in boom width should be considered alongside field size and shape as a wide boom in a small awkward-shaped field can result in driving problems.

Tank

A large tank reduces the proportion of time spent travelling to refill the sprayer in the spraying cycle. Large tanks are more useful when used in conjunction with wide booms, as the larger tank can be used to advantage with the faster rate of emptying of the wide boom. A larger tank means greater weight and therefore soil compaction. To help reduce soil compaction, farmers should consider the sprayer type (e.g. mounted, trailed or self-propelled) and the size of tyres fitted to the sprayer. The greater the sprayer weight, the larger the tractor required to pull or carry the sprayer. The use of a tractor front-mounted nurse tank will give better weight distribution with an increased spraying capacity.

Availability of water

The output of the spraying system can be greatly increased if water is in close proximity to the field being sprayed. The availability of water from ponds, dykes and rivers should be taken into consideration, particularly if the sprayer has a self-fill water hose. When spraying fields close to the farmstead, the sprayer can return to the farm, provided that a fast rate of fill can be obtained. Large overhead tanks can be used to speed up sprayer filling; the tanks fill up from the mains over a period of time and empty via gravity. An alternative method can be the use of large, inexpensive, former orange juice containers and a petrol-driven centrifugal pump to provide a fast rate of fill.

Water bowsers

The use of a water bowser will increase sprayer output dramatically because the time spent travelling to the water source is considerably reduced. A simple water bowser can comprise of a number of old clean tanks fitted on a farm trailer; the tanks can be emptied via the sprayer self-fill hose or a centrifugal pump. Centrifugal pumps can give a high rate of fill (e.g. 450l/min) but must be situated below the tanks as they need to be primed. Large bore pipes will allow high flow rates. Ideally a water bowser should be situated in the field, to meet the sprayer at the end of a run, thus avoiding the need for the sprayer to return to the field gate and the resultant time-consuming folding of the boom. Access to the field may be limited by soil type and moisture content so the bowser could have a long pipe on a gantry allowing the sprayer to refill while the bowser stays outside the field.

Pumps

Centrifugal pumps transfer liquids at high speed and low pressure and are ideal for filling sprayers or water bowsers. The pump may be driven by a small petrol engine; good pre-season maintenance will ensure the engine starts! On some larger sprayers, centrifugal pumps for filling the sprayer and agitating tank contents are fitted as standard. Filling at

450 litres per minute leads to a rapid turnaround of the sprayer, making it possible to take full advantage of good spraying conditions.

Mixer tanks
The mixer tank is the next stage to consider after water bowsers. A water bowser may fill a sprayer quickly but it may still take a long time to mix pesticides with water. Mixer tanks – basically a tank with a centrifugal pump and large bore pipes mounted on a trailer – combine the advantages of a water bowser with a very good agitation or mixing system. When a second person is employed, a very high output can be achieved, as the sprayer driver can remain on the tractor/sprayer and the second person can be in charge of measuring chemicals, mixing and filling. The modern trend towards a chemical fill probe or a mixing bowl ensures safety because the operator doesn't have to clamber over the sprayer with a container of neat pesticide. On large fields the mixer tank can discharge mixed spray straight into the sprayer as it is already mixed and doesn't require measuring out.

Forward speed
Forward speed will affect the output of the spraying system. Forward speed depends on the nozzles selected, ground conditions and boom stability. Undulating or rough ground causes boom bounce and yaw resulting in uneven application. A growing number of farmers use an automatic spray volume control which allows the sprayer to maintain a constant output at varying forward speeds. Forward speed whilst travelling for water is also important; if farm roads or tracks are in good condition then high road speeds can be maintained. Certain tractors (e.g. JCB Fastrac or M.B. Trac) have very high forward speeds and with a rear platform are attractive sprayer tractors, particularly on the larger farm or estate.

Application rate
The introduction of low volume application (e.g. 75–100l/ha) drastically reduces the amount of water required. A tank of a given volume will allow for longer periods of spraying and therefore reduce the amount of time spent travelling to refill and refilling.

The operator
The sprayer operator who is responsible, well trained and highly motivated, is the key to achieving a high output from the spraying system. Communication is most important so farmers should explain how the chemical works (e.g. if the chemical action is systemic or contact) and outline any problems which may arise during application such as drift. Some farmers issue job sheets outlining exactly what is required per field. Field records need to be comprehensive but straightforward. Some farmers have radio communication systems and these can help as follows:

● if a breakdown occurs, the operator can summon assistance quickly thus reducing the time spent on repairing the sprayer or tractor. This is sometimes referred to as reducing 'down-time'
  ● if a query arises regarding the use of a chemical, then a second opinion can be quickly sought
  ● similarly, if the weather looks changeable, then advice can be sought
  ● safety – radio communication is most useful for the single operator, if an accident should occur, help can be summoned quickly

The availability of labour is important: if a second person is available, they can be employed to ferry the water bowser or operate the mixer tank. Supplying a mealtime relief driver will assist operator safety because of the reduced risk of food contamination. A rest period will help maintain sprayer output. A safety-conscious employee, provided with a safe working environment, will help maintain a high output from the spraying system.

The field

Field size, shape, position and topography will affect system output. Small irregular-shaped and undulating fields will reduce output considerably and obstructions within the field (e.g. trees, telegraph poles, etc) will have the same effect. Wide booms require wide headlands on which to turn if output is to be maintained. The distance of the field from a water source will affect output, particularly if it is not possible to use a water bowser.

Forward planning the sequence of field spraying will help maintain a high output; spraying fields in sequence with the same product(s) avoids many hold-ups, e.g. for de-contaminating the sprayer.

MAKING QUALITY SILAGE – MANAGING A GRASS HARVESTING SYSTEM

Pre-season preparation is the key to ensuring a smooth harvesting system.

Tractors

Check the correct operation of brakes, clutch adjustment, tyres and ballast. Repair any oil/fuel leaks. Carry out a thorough service. Surveys of tractors used with forage harvesters indicated that the most common fault was poor maintenance, in particular dirty air cleaners, old injectors and poorly calibrated injection pumps. Cleaning an air filter is such a simple task to perform yet is often overlooked. A dynamometer test from a local machinery dealer will indicate the power available compared to the OECD-rated power.

Grass roller
Grease the bearings and fill with water if necessary.

Mower/conditioner
Check the blades, belts, gearbox oil level, stone guards and pto guard.

Forage harvester
Check the pick-up tines are straight, gearbox oil level, belts/chains aren't stretched, the condition of the cutting cylinder knives, feed rollers, hydraulic and electrical connections and controls. Ensure the correct shear bolts are fitted and electrical controls function properly.

Additive applicator
Using water, check the electric pump works, check pipes for leaks, and ensure electrical connections aren't corroded. Check the flow rate.

Trailers
Check the brakes operate, the hydraulic hoses aren't leaking or chafed and the connectors don't leak. Ensure the tailgate operates smoothly, wheel nuts are tight, tyres are inflated and the wheel bearings are set correctly.

Buck rake
Check the tines are straight, hydraulic pipes aren't worn/leaking, connectors, and that the push-off ram operates and doesn't leak.

Fertiliser spreader
Check the drive mechanism, belts, chains, etc, the delivery and control mechanism functions properly and the wheels/tyres are in good order.

Roads and tracks
Fill potholes because they damage trailer wheel bearings and slow down tractors. Trim hedges back near gateways to ensure good visibility (for tractor driver and other road users). Hang gates so that they open easily.

Emergencies
Know where the bottle jack is kept! Decide on puncture repair policy – farm or tyre depot. Keep a list of emergency telephone numbers.

Matching Windrows To Tractor Power, Gear Ratios And Topography

It is crucial that the harvester reaches its optimum output. A number of inter-related factors affect harvester throughput including:
- swath width and its resultant windrow size
- tractor power and gear ratio
- tractor, harvester and loaded trailer weight
- in-line or side loading

● terrain

The harvesting combination in Table 5.2 is able to tackle the full range of windrows because of its engine power and gear ratios. Note that there is a lower output in a 3m windrow than a 2.5m swath (due to gear ratios)

Table 5.2 SPOT WORKRATE (TONNES/HR) FOR A 75KW TRACTOR WORKING ON LEVEL GROUND					
Method of loading	Swath/window spacing (m)				
	1.5	2.0	2.5	3.0	4.0
In-line trailer	33	34	41	39	41
Side-loading trailer	33	34	43	39	41

Tractor 75kW (100hp): weight 6.8 tonnes, Forage harvester weight: 2.0 tonnes
Trailer loaded weight: 7.1 tonnes

and that a 2.5m swath gives the same spot throughput as a 4m swath when towing a trailer in-line, but there will be more rows/ha, and therefore more travelling resulting in a lower overall workrate! For the same reason, and because the tractor is relatively heavy for its rated output, the benefit of side loading is not so marked. Other tractor/forage harvester/windrow/power combinations have different affects on output. Pulling a trailer in-line with a wide swath on a hill may cause the driver to change down a gear, thereby losing throughput.

Field trial research shows that many forage harvesters are operated at relatively low forward speeds, perhaps due to concerns about damage caused by stones or hitting the ground. If this is the case, there is scope to utilise more of the swath width created by mowers/conditioners. Side loading, whilst requiring an extra person and tractor/trailer, often produces an increase in output of between 10% and 20%, depending on power, weight, terrain and swath width.

Forage harvester design also affects power requirement and therefore output. Reverse cut foragers with one or no feed rollers require less power than a precision chop forager with six feed rollers. Foragers with flywheels run smoothly in lumpy swaths and have fewer variations in power demand.

Analysing A Grass Harvesting System

The observations made in the section analysing a grain harvesting system are equally applicable to a grass harvesting system. Output is dependent on many variables and a systems analysis approach is required. The major sections to be analysed include factors which affect:
● system efficiency
● biological factors
● machine design
● environmental factors

- financial factors
- time available

Falling 'D' or digestibility values in the grass, area to be harvested and management/staff motivation are similar to the ripening/shedding conditions found in a cereal harvest.

The costs of owning a forage harvester (and the large tractor required to pull it) can be calculated by using the computer programme found in Appendix B and described in Chapter Seven.

The use of a partial budget and break-even analysis, described in Chapter Seven, can be used to decide if consideration should be given to employing a contractor rather than owning a forage harvester. As with the example given with a combine harvester, the area of grass required to justify such a financial investment is quite large.

THE INTRODUCTION OF REVOLUTIONARY TECHNOLOGY – A CASE STUDY ON THE INTRODUCTION OF A GRAIN STRIPPER HEADER

The following section considers the purchase of a grain stripper header for the combine at Acorn Farm. The stripper header, developed at Silsoe Research Institute and sold by Shelbourne Reynolds, has revolutionised harvesting in certain crops and conditions.

The introduction of such an attachment has many repercussions throughout the harvesting system. In order to decide whether to buy a header the farmer must conduct a thorough background research and consider the advantages claimed by independent sources. These include the following:

- research shows a reduction in combine header (table) losses in barley, linseed and peas but an increase in header losses in wheat, oats, beans and oilseed rape
- better timeliness and an increase in throughput, e.g. winter and spring barley

As an average of all cereals there is an increase in throughput by 50% with a conventional combine but an increase of 90% with a multi-cylinder combine. Better timeliness will allow:

i. easier cultivations, carried out in better conditions

ii. better drilling conditions and, perhaps, more even sowing

iii. better crop response, more even germination

The adoption of a grain stripper header may give a positive long-term effect (e.g. a stripper header operating at 5mph and a cutterbar header operating at 3mph with a 200-hour harvest period with a conventional cutterbar combine will result in 80 hours less time with a stripper header).

Each hour saved is money saved but the main question is how much is timeliness worth. How do you quantify timeliness? Good conditions for fieldwork will result in a higher output and less fuel. Better crop

response will, hopefully, result in better yields, therefore improving income. The advantages are:
- fuel saving: Shelbourne Reynolds claim a 25%+ saving due to improved speed/ha and logistics
- harvesting time: should be earlier and dryer capacity improved.
- long-term developments: the design of combines may lead to smaller, cheaper units

The disadvantages are:
- capital cost: should a stripper header be purchased as an addition for an existing cutterbar combine or purchased instead of a cutterbar system? This will depend on the range of crops grown
- repairs and maintenance costs: two rows of stripping combs need to be replaced every 240–320ha
- straw: as the stripper header removes the ears there is a large amount of standing straw. This can be dealt with by:

i. using an existing plough but it should have good under-beam clearance and the soil type and its effect on rotting down straw should be considered

ii. using an existing set of rolls/press on a tractor

iii. using an existing set of discs but this creates an extra cost of one-two passes

iv. buying a tractor-trailed windrower plus running costs

- power: a stripper header needs four times more power than a cutterbar header but less power for threshing and separation (on average an 8% increase in power)

Remember: An increase in output/hr, therefore total power/ha or power/season may be less than with a cutterbar header
- increased output results in:

i. a requirement for more trailer(s), staff and tractors

ii. their organisation

iii. perhaps a larger grain reception pit

iv. possible soil damage from larger trailers if their wheels are too small

- stubble cleaning. How will this be done?
- potential straw breakdown in the soil

After considering all the above the farmer must then look at the positive/negative aspects via the construction of a partial budget (as described in Chapter Seven). What will be the knock-on benefit or the effect of long-term changes to the whole farm system?

Further reading
How to choose and use combines. Publ. No.88. SAC, Edinburgh.

Audsley, E. 1989. *The value of a combine harvester hour*. Divisional Note DN1554. AFRC Engineering, Silsoe: SRI.

Hale, O. 1989. *The grain stripper harvester*. Silsoe: SRI.

CHAPTER 6

MACHINERY SELECTION

Before obtaining farm machinery, farmers need to address a number of very important questions concerning the most appropriate system for their farm.

- what is the field operation requirement?
- are there any time constraints?
- what type of machine is required?
- what size and how many?
- who will operate the equipment?
- when is the best time to replace?

A number of inter-related factors influence the selection of farm machinery and follow the systems analysis guidelines laid down in Chapter Two. They include:

- management objectives, circumstances and attitudes
- environmental factors: soils, climate, topography, field sizes and locations, fieldwork days, compaction risk, legal constraints
- labour availability: quantity, skills, aptitude
- crop and livestock factors: operational requirements, timeliness aspects, materials handling
- cropping pattern factors: scheduling, peak workloads, and capacity requirements
- machine and system performance: workrates, power requirements and performance, suitability, flexibility in use
- costs: capital costs, depreciation, repairs, fuel
- acquisition and financing: ownership, lease, hire purchase, contract hire, contractor, new or secondhand, taxation
- institutional support: dealership facilities, workshops, spares and services

Farm machinery selection must take into consideration the seasonality and unpredictability of the amount of time available for fieldwork. A step-by-step approach is recommended:

- identify operations and their peak work periods
- schedule and sequence operations within peak periods
- estimate available fieldwork days.
- estimate machinery capacity requirement (ha/hr or ha/day) to perform the task
- select machinery types, sizes and numbers

Ideally the whole farming calendar and its constituent operations should be considered simultaneously.

TRACTOR SELECTION

The purchase of a tractor is one of the most important decisions to be made on any farm. There is a vast array of makes, models and sizes with a myriad of optional extras available to the farmer. A correct decision will benefit the business considerably but the wrong decision will be an expensive mistake to be regretted for many years.

What Make And Model Of Tractor?

There are many factors to be considered before purchasing a tractor. On many farms the investment in a tractor may be the largest capital outlay after the purchase of the land.

The type of tractor is determined by:
- existing and future farm policy: farm size, cropping, labour
- existing and future tractors/equipment: tractor use, soil type, power requirements, implement widths
- replacement policy
- cost: capital cost, spares and repairs, potential resale value
- manufacturer: reputation, reliability, warranty conditions, financial inducements
- equipment dealer: proximity, availability of spares, reputation and rapport, sales/service record, skilled workshop staff, cost of spares and service, trade-in deal
- preference of the operator
- preference/prejudice of the purchaser
- the tractor: basic specifications/options/extras, engine power, pto/draught power, fuel consumption/economy, fuel tank capacity, transmission; synchromesh versus basic number of gears; change-on-the move torque converter (road speed), four-wheel versus two-wheel drive; tyres: radial/cross-ply, sizes; hydraulics: flow rate, number of spool valves, lift capacity/assister rams; controls: ease of use, trailer brakes, weight, basic weight; availability of front/wheel weights, ease of attachment of weights; dimensions: turning circle; height (cab clearance) and width; cab layout/ease of use of controls, visibility, air conditioning, seat/comfort, ease of on-farm-maintenance

Power
Power is measured in watts (W). 1 watt = a force of 1 newton (N) acting through a distance of 1 metre in 1 second.

$$1 \text{ kW} = 1000 \text{ watts} = \frac{Nm/sec}{1000}$$

Power = torque multiplied by speed
Force is measured in newtons (N)

1 newton = the force which produces an acceleration of 1 metre per second when acting on a mass of 1 kilogram

$$1 \text{ kN} \quad = \quad 1000 \text{ newton} \quad = \quad \frac{kgm/s^2}{1000}$$

Drawbar power
Drawbar power is measured in kilowatts (kW). Drawbar power = pull multiplied by speed

$$kW \quad = \quad \frac{draft \ (kN) \ x \ speed \ of \ travel \ (m/h)}{60 \ (min/hr) \ x \ 60 \ (sec/min)}$$

Table 6.1 TRACTOR HOURS (per annum)		
Crops		
	Per hectare	
	Average	**Premium**
Cereals	9	7
Straw harvesting	3.5	2.5
Potatoes	25	15
Sugar beet	20	12
Vining peas	20	12
Dried peas	10	8
Field beans	9	7
Oilseed rape	9	7
Herbage seeds:		
1 year undersown or 3 year direct drilled	7	5
1 year direct drilled	11	8
Hops (machine picked)	125	-
Kale (grazed)	8	6
Turnips/swedes: folded/lifted	12/35	10/25
Mangolds	50	35
Fallow	12	7
Ley establishment		
Undersown	2	1
Direct seed	7	4
Making hay	12	8
Making silage		
1st cut	12	8
2nd cut	9	6
Grazing:		
Temporary grass	3	2.5
Permanent grass	2	1.5

Table 6.1 TRACTOR HOURS (per annum) - *continued*	
Livestock	
	Per head Average
Dairy cows	6
Other cattle over 2 years	5
Other cattle 1–2 years	4
Other cattle 0.5–1 year	2.25
Calves 0–0.5/2 year	2.25
Yarded bullocks	3
Sheep, per ewe	1.25
Store sheep	0.8
Sows	1.75
Other pigs over 2 months	1
Laying birds	0.04

Notes

1. Sources include The Farm as a Business: Aids to Management, 6: Labour and Machinery. *'Premium' figures apply where high performance machinery is used.*

2. For livestock, annual requirements are the per head requirements above multiplied by average numbers during the year (i.e. average numbers at end of each month).

3. As with labour, the number of tractors required by a farm depends more on the seasonal requirements and number required at any one time than on total annual tractor hours. These can be calculated from the seasonal labour data provided in Nix's booklet. The soil type and size/power of tractors purchased are obviously other relevant tractors.

Source: *Farm Management Pocketbook*, John Nix, 26th edition September 1995, Wye College

As most farmers still refer to tractors by engine or pto horsepower, the following conversions will be useful:

1 horsepower	=	0.746 kW
kW	=	hp multiplied by 1.341
hp	=	kW multiplied by 0.746

Table 6.1 shows 'standard tractor hours', a system first devised by Sturrock (1957) after observing tractors working in East Anglia. The system was subsequently developed by Culpin (1959) and is published in the ubiquitous *Farm Management Handbook* by Nix (2000).

Example

The labour and machinery profile for Acorn Farm was shown in Chapter Three, Figure 3.6. Acorn Farm comprises:

Spring barley, 100ha
Winter barley, 100ha
Winter wheat, 150ha
Oilseed rape, 58ha
Potatoes, 35ha
Grass
Undersown/year: 6ha,
Hay, 10ha
Silage
First cut, 30ha
Second cut, 30ha
Grazing – permanent pasture 30ha, temporary ley 20ha
Dairy cows: 120
Followers, over 2 years: 20
Followers, 1-2 years: 25
Followers, 1/2-1 year: 20
Calves, 0- 1/2 year: 25

Using the data provided in Table 6.1, the standard tractor hours of
Table 6.2 can be estimated. If an average medium-sized tractor works

Table 6.2 STANDARD TRACTOR HOURS FOR ACORN FARM

Enterprise	Hectares or head	Standard tractor hours/ha or/head	Total
Spring barley	100	9	900
Winter barley	100	9	900
Winter wheat	150	9	1350
Oilseed rape	58	9	522
Potatoes	35	25	875
Grass			
Undersown	6	2	12
Hay	10	12	120
Silage 1st cut	30	12	360
Silage 2nd cut	30	9	270
Grazing			
Permanent pasture	30	2	60
Temporary ley	20	3	60
Dairy cows	120	6	720
Followers, > 2 years	20	5	100
Followers, 1-2 years	25	4	100
Followers, 0.5-1 year	20	2.25	45
Calves, 0-0.5	25	2.25	56.25
		Standard tractor hours total	16450.25

800 hours per year, the number of medium-sized tractors required at Acorn Farm is

$$\frac{6450.25}{800} = 8$$

This method is a fast, simple system, which occasionally provides an approximate indication of the number of tractors, but has a number of limitations. The major disadvantage is that there is no indication of number and size of tractor. What size is a small/medium tractor? Instead of using 12 small tractors, could 6 large tractors be used? What is the relationship between small and large tractors? What is the basis for choosing an eight wheel giant tractor? How is the pto hp requirement for a forage harvester taken into consideration?

What Size Of Tractor Should Be Used For Draught Work?

Soil type and cultivation
A good method of tractor selection, based upon scientific fact, is to consider soil type and the draught requirement necessary for different types of cultivation equipment.

To decide what size of tractor will match the existing draught equipment on the farm the following steps should be taken.

Step one
Construct a labour/machinery profile. This will provide the cropping schedule, time available, the number of people required and the number of tractors (assuming each person has one tractor).

Step two
Decide upon the output of the implements within the time, or approximate time available (e.g. 2-week timeframe to plough the fields).

Either refer to a farm management handbook: where a plough rate is x ha/hr (adjust output to your soil, your local knowledge), or use first principles such as speed x width x efficiency divided by ten.

Step three
Determine the physical characteristics of the soil, leading to the draught requirement (Table 6.3).
- establish width and speed of implement
- establish soil draught characteristics from Table 6.3

e.g.
i. ploughing a firm clay soil with a 1.4m plough at 9km/hr
ii. a heavy soil has a draught of 18kN/metre
Total draught = 1.4m x 18kN = 25.2kN

Table 6.3 TYPICAL DRAUGHT REQUIREMENTS

Implement or machine	Typical forward speed (km/h)	Draught (per m width) Heavy	Medium Light	pto power (per m width)
Mouldboard plough	5–6.5	18kN	12.5kN 6kn	-
Mole plough	3–3.5kN	Up to 44kN/mole		-
Sub-soiler	4–4.5	Up to 17kN/tine		-
Disc harrow	8–10	6KN	3.5kN 1.5kN	-
Medium-tine harrow	8–10	9kN	3.5kN 2kN	-
Spring-tine harrow	8–10	7kN	2.5kN 1.5kN	-
Rotary cultivator (e.g. Rotavator)	4–4.5	From 0 to 1.85kW		24.5kW
Reciprocating harrow	5–6.5	Approx 2.45kW		7.35kW
Rotary harrow (e.g. Roterra, Maschio)	5–5.5	Approx 2.4kW		12.25kW

Source: Davies, Eagle and Finney (1993)

Step four

Use the formula to determine drawbar power and adjust for soil condition (typical forward speeds may be found in Table 4.4).

$$\text{Drawbar power} = \frac{\text{total draught (kN) x speed (km/hr)}}{3.6}$$

$$= \frac{25.2 \times 9}{3.6} = 63 \text{ kW}$$

Remember this is drawbar power. A tractor loses approximately 50% in transmission/power losses.

Power loss occurs due to mechanical inefficiency in the gearbox and back axle, drive to hydraulic pumps, wheel slip and rolling resistance.

To convert drawbar power to engine power, see Table 6.4 on available tractor power with varying soil conditions. A firm soil allows about 60% of engine power to be utilised at the drawbar:

$$63 \text{ kW} \quad \times \frac{100}{60} = 105\text{kW}$$

Table 6.4 AVAILABLE TRACTOR POWER WITH VARYING SOIL CONDITIONS	
Power source	Per cent
Engine	100
pto	86
Drawbar (firm soil)	60
Drawbar (tilled soil)	50
Drawbar (soft soil)	45

A tractor with approximately 105kW at the engine should be considered if all the ploughing is on firm, heavy soils.

At Acorn Farm, a soft soil is often found when ploughing land after a potato harvest. Application of the data in Table 6.4 shows that a soft soil allows 41% of engine power to be used due to the increase in rolling resistance of the tyres (imagine the tyres sinking into the soft soil and needing more power to 'climb up and out' of the wheel mark compared to firm soil).

What Size Of Tractor For pto Work?

Not all farms need to size their tractors on the maximum draught requirements such as ploughs or subsoilers. Many livestock farms need to consider pto power for forage harvesters or powered rotary harrows. Follow the example above using pto power instead of draught power requirements.

The power requirement for a forage harvester varies considerably, according to:
● forager design (e.g. flywheel versus reverse chop versus metered chop)
● the towing of a trailer behind the forager
● terrain
● condition of the tractor

Chapter Five reviews grass harvesting systems and gives information on output according to matching swath width and gear ratios. Tractor power survey results in grass harvesting areas are also noted.

What size implement to match existing tractor power?
The following formula can be used to match a given pulling ability:

$$\frac{\text{available draught (kN)}}{\text{draught/metre of implement (kN)}} = m$$

At Acorn Farm, one of the tractors has 30kN of available draught at the drawbar, what width of disc harrow could it pull? Table 6.3 suggests that discs have a draught of 6kN/metre width.

e.g. $\dfrac{30kN}{6kN}$ =   5m

Tractor Power Surveys (Table 6.5)
Make an inventory of your tractors, excluding self-propelled sprayers but including fork-lift trucks and measure the power against the land available.

e.g. $\dfrac{112kW}{30\ ha}$ =   3.7kW/ha

Table 6.5 TRACTOR POWER SURVEYS		
Survey area	kW/ha	hp/acre
A. Main cultivation tractors (East Anglia: arable)	1.1–1.3	0.6–0.7
B. Main cultivation tractors (Midlands: arable)	1.1	0.6
C. All farm tractors (U.K: mixed farms) Average :	0.75–5.6 1.87	0.4–3.0 1
D. Main cultivation tractors (Large scale arable farms)	0.84	0.45

Torque
Torque, the rotating force or pull, is the product of a force (measured in Newtons) and the radius of the shaft (measured in metres). Torque is measured in Newton metres (Nm).

When purchasing a tractor it is as important to know the torque reserve of an engine as it is the power. The torque reserve (sometimes called torque back-up, or torque rise) is the ability of the engine to withstand changes in torque loads, such as when a great clump of grass goes through a forage harvester. The same effect can be seen on a combine harvester when a wad of straw goes through the drum/concave. Ideally the torque reserve should be as high as possible (e.g. in excess of 15%).

How is torque measured?
The European organisation OECD has laid down a standard code for testing tractors. All new tractor models are independently tested at

centres across Europe which provide many independent test reports giving detailed information on tractor performance. The test includes information about power at the engine, pto, at the drawbar, hydraulic performance, noise, centre of gravity, braking and turning circles. Another standard which is used for testing an installed engine is DIN 70020. SAE (USA) records the maximum power an engine achieves. The specifications of a tractor, including important information regarding engine performance, is usually located in the tractor brochure. Detailed results of OECD tractor tests are to be found in the excellent farm machinery magazine, *Profi* or the OECD web page (see web sites).

Figure 6.1 indicates that maximum torque of the New Holland 6640 occurs at 1200 engine rpm, whereas maximum power occurs at 2200 engine rpm. If the tractor is ploughing a field the tractor pull will keep increasing for 1000rpm. If the plough hits a 'sticky patch of heavy clay' the torque reserve will allow the engine speed to reduce by 1000rpm. With the New Holland 8240 the maximum torque occurs at 1200 engine rpm and maximum power occurs at 2100 engine rpm.

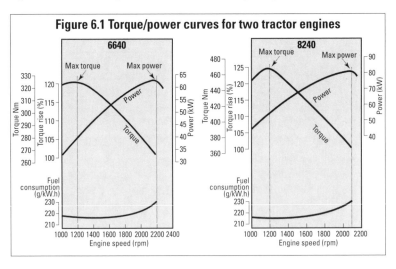

Figure 6.1 Torque/power curves for two tractor engines

When a tractor works at a faster engine speed you can make use of the higher power output. As the tractor meets an additional load, the engine slows down and the decrease in speed brings an increase in torque.

% torque rise $\quad = \quad \dfrac{\text{maximum torque} - \text{rated torque}}{\text{rated torque}} \times 100$

New Holland 6640:

$$\frac{327 - 270}{270} \times 100 \quad = \quad 21\%$$

New Holland 8240:

$$\frac{465 - 370}{370} \times 100 \quad = \quad 26\%$$

With similar sized engines, maximum torque should occur at 65-75% of the engine rpm occurring at maximum power. A recent survey of tractor engines shows a range of torque rise (reserve) of 29–53% (Table 6.6).

Make & model	Engine power kW	Max. torque Nm	Torque rise/ engine speed drop
New Holland 6635	63.0@2500rpm	288@1400rpm	29.7/44.0%
New Holland 7740	65.8@2050rpm	385@1400rpm	29.2/33.3%
MF 4255	69.8@2200rpm	364@1500rpm	30.0/31.7%
Case MX135	99.0@2200rpm	521@1600rpm	32.6/27.3%
Fendt 515C	110@2300rpm	574@1300rpm	39.0/43.5%
John Deere 7710	114@2100rpm	725@1200rpm	53.0/42.9%

Table 6.6 TORQUE RISE ON MODERN TRACTORS

Source: Compiled from OECD test data at the DLG test station

Speed range

Speed range is a measure of the difference between the rated speed and the stalling speed. A combination of good torque reserve and good speed range, as seen in the tractors in Table 6.6, shows that those tractors have good, flexible engines, allowing the tractor to work in a range of conditions.

$$\text{Speed range} \quad = \quad \frac{\text{rated speed} - \text{stalling speed}}{\text{rated speed}} \times 100$$

e.g. Case MX 135

$$\text{Speed range} \quad = \quad \frac{2200 - 1600}{2200} \times 100 \quad = \quad 27.3\%$$

Tyres And Traction

Many farmers buy a quality tractor but sadly forget to specify one of the most important areas of the tractor, the tyres. Often, the range of tyres (Figure 6.2) offered for a tractor by its manufacturer is limited, but tyre dealers and tyre companies can often supply alternative tyres or wheel equipment. Wheels and tyres must be selected and maintained carefully to decrease fuel consumption, prolong tyre life, reduce soil compaction and improve workrates.

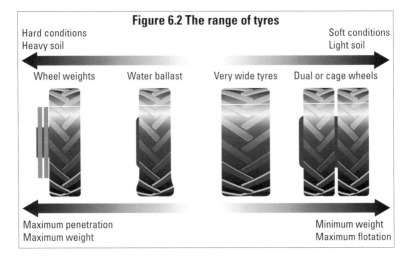

Figure 6.2 The range of tyres

Hard conditions
Heavy soil

Soft conditions
Light soil

Wheel weights   Water ballast   Very wide tyres   Dual or cage wheels

Maximum penetration
Maximum weight

Minimum weight
Maximum flotation

Factors to be considered include:
- the likely load upon the tyres, single versus dual wheels
- soil conditions such as moisture content and risk of compaction
- working environment such as the extra stability provided by four-wheel drive, and tyre width for working in row crops
- standard of tyre manufacture, radial versus cross-ply, operating pressure and tread pattern.
- cost

The following should also be taken into account. The area of tyre in contact with the soil affects flotation, rolling resistance, ground pressure and frictional grip. The load on the tyre affects ground pressure, frictional grip, penetration and rolling resistance. The soil type affects flotation, sheer strength and penetration.

Fitting a large wide single tyre is an alternative to duals, but duals offer the advantage that the outer wheels can easily be removed when not required, thus providing greater flexibility. Fitting dual wheels will reduce rolling resistance and increase tyre contact area.

If the tractor is to operate on predominantly hard surfaces, then a smaller tyre at high inflation pressure will be satisfactory. Conversely, if the tractor is used mainly on softer soils where sinkage and soil compaction are likely to be a problem, a larger tyre operating at the minimum inflation pressure or dual wheels is recommended. On hard surfaces, use high inflation pressures to reduce wear and on soft soils, low inflation pressures to reduce rolling resistance and soil compaction. It is preferable to increase tyre diameter rather than tyre width when contemplating larger tyres, but often tyre diameter is restricted by the design of the tractor.

COMBINE HARVESTER SELECTION

The method of calculating workrate, described in Chapter Three, using speed multiplied by width multiplied by efficiency, divided by ten, cannot be used in the selection of a combine harvester. The output of a combine is dependent on factors other than cutterbar width and most combine models are available with a number of cutterbar width options.

Combine throughput is dependent on a number of mechanical and operational factors.

Mechanical Factors Affecting Output

Threshing drum and concave size

The majority of threshing (85%) takes place as the crop passes through the drum and concave provided the combine is set up properly. Throughput will obviously increase if the threshing area is as large as possible. High output combines have wide drums with a large concave wrap.

Straw walker area

Traditionally, straw walker area was used to determine the output of a combine. The larger the walker area, the greater the opportunity to shake out the ears and grains trapped in the straw and so an acceptable level of grain loss determined the combine output. Secondary separation devices are found on most modern combines and improve throughput whilst reducing grain loss. Although the straw walker area is important, it is only part of the decision tree. For example, the Case CF80 has 7.40m^2 of straw walker area.

Sieving area

The larger the sieving area, the greater the opportunity to clean and grade the grains. If the sieving area is too small, excess grains will be lost over the back of the sieves. There is little point in having a large threshing and walker area if the sieving area limits throughput. For example, the Case CF80 has 6.25m^2 of sieving area.

Breakdowns

Buying a quality combine harvester is important. Good preventative maintenance will minimise breakdowns. Allowing trained operators time to maintain their combines is equally important. Cleaning radiators in dusty conditions will ensure engines don't overheat and minimise downtime. A number of farmers hire air compressors specifically for cleaning combines during the harvesting season. A well-trained operator will ensure good maintenance and check drive belts to ensure they don't break during ideal harvesting conditions.

Operational Factors Affecting Output

Driver skill

Keeping the cutterbar full, matching throughput with acceptable grain loss is the key responsibility of the driver. The driver's ability and attitude is important.

Forward speed

Driving a combine at speeds over 5km/hr requires great skill; most operators would prefer to drive slower and use a wider cutterbar. Harvesting fields where stones can be a problem requires attention, therefore slower speeds are a must. Table 6.7 shows the forward speed required to maintain 10t/hr throughput.

Table 6.7 FORWARD SPEED NEEDED FOR 10T/HR STRAW INTAKE IN KM/HR				
	Effective width of cut			
Straw yield (t/ha)	3.5m	4.0m	4.5m	5.0m
4.5	6.3	5.5	4.9	4.4
6.0	4.8	4.2	3.7	3.3

Source: SAC 1982

Management expertise and support

A high level of farm management (and good weather) will ensure the combine harvester is presented with ripe grains in ideal harvesting conditions. Farm management will also include good logistical support such as labour and trailer organisation, as described in Chapter 5. Harvester and trailer driver relief over lunchtime will improve output. Good team effort is one of the major factors which can improve output. Management skills can be honed via training and attitude must always be positive.

Blockages

Good husbandry will ensure a clean, standing crop thus avoiding laid corn as a result of excessive fertiliser use. A good, timely spraying programme will reduce weed-ridden crops. The cropping programme such as the spread of crops (e.g. winter-/spring-sown crops, winter barley to beans) will result in a longer harvesting season.

Moisture content of the grain and straw

An important biological and management factor. The decision as to when to harvest the crop depends on many related factors, but drying and storing facilities, area to harvest and attitude to risk must be considered.

Weather
The vagaries of a Northern European summer must be considered!

Farm layout
Factors such as shape and size of field, distance between fields and terrain will also affect output.

Assessing The Seasonal Capacity Of Conventional (Straw Walker) Combines

As mentioned above there are a number of mechanical and operational factors which affect output. MAFF (1969) conducted a survey of combine outputs; their method is quite old but it still offers a good guide. They suggested the following formula for sizing combines:

$$A = \frac{C\,W\,P\,H\,U}{Y}$$

A = area harvested per season (ha)
C = Constant 2 .7
W = Straw walker area (m²)
P = Performance factor (Table 6.8)
H = Length of season (hours)
U = Utilisation factor (management) (Table 6.9)
Y = Average yield of grain (tonne/ha)

The number of hours available for combining depends on the weather in the area of the country and the type of crops to be harvested. Actual hours available will vary from 100 to 300. Acceptable moisture content

Table 6.8 COMBINE DRIVER'S ABILITY	
Performance factor	**(driver's ability) (P)**
Very high	0.65
High	0.6
Above average	0.55
Average	0.5

Table 6.9 UTILISATION FACTOR (MANAGEMENT) (U)				
Standard of Management	**Combine season, hours**			
	< 150	**151-200**	**201-250**	**250+**
Very good	0.75	0.7	0.65	0.6
Good	0.7	0.65	0.6	0.55
Average	0.65	0.6	0.55	0.5

and dryer availability/capacity must be taken into account, as must the spread of crops and ripening.

Example

How many hectares will Model X combine harvest in a given season? Background information:

Straw walker area = 7m^2
Length of season = 200 hours
Performance factor (driver's ability) = 0.65
Utilisation factor (management) = 0.7
Average yield of grain = 8t/ha

$$A = \frac{C \times W \times P \times H \times U}{Y}$$

$$A = \frac{2.7 \times 7 \times 0.65 \times 200 \times 0.7}{8} = \frac{1720}{8} = 215\text{ha}$$

If a farmer wishes to know what size combine (straw walker area) to obtain the following formula may be used:

$$W = \frac{A \times Y}{C \times H \times U \times P}$$

How many hours are available:

$$H = \frac{A \times Y}{C \times W \times U \times P}$$

If a farmer uses published workrates as a guide to buying a combine, he should be wary of advertised claims of high output. Note the difference between spot workrate and seasonal workrate in Table 6.10. With modern 'rotary' combines it is hard to assess the seasonal capacity. Take the manufacturer's claimed output and compare it with a neighbouring farm output and a trial on your farm.

Table 6.10 AVERAGE COMBINE WORKRATES

SPOT	OVERALL	SEASONAL
	Medium size combine	
Ha/hr t/hr	ha/hr t/hr	ha/hr t/hr
1.67 8.4	1.29 6.5	1.14 5.7

Source: SAC 1982

Replacing Two Small/Medium Combines With One Large Machine

Benefits
- there may be a lower capital cost as one large machine may cost less than two smaller combines of the same total capacity
- one main operator required means savings in labour costs as there is probably one fewer person to employ
- good for staff morale, especially the potential driver
- lower variable costs such as repairs, maintenance, fuel and labour
- improvement in timeliness of operations

Disadvantages
- medium-size combines may have a better trade-in as they fall into a larger market sector but this depends on region and export markets
- loss of harvesting flexibility
- two smaller machines might provide a larger cutting width and have a greater drum/concave width
- consequences of a mechanical breakdown upon output and timeliness
- physical size of the machine could be a transport problem in certain fields and on certain roads
- illness of a trained operator could have a major impact on output

CROP SPRAYER SELECTION

Background Considerations

Existing and future farm policy and equipment
Existing and future farm policy will dictate the area, variety and rotation of the crops to be sprayed; different crops will have different spraying requirements, such as types of chemical, application rates and the timing of applications.

If existing or future cropping policy includes root crops, then the sprayer boom width and wheelbase will need to match row or bed widths to ensure the correct operating width and to prevent damage to the row crops by the tyres.

Future policy changes will affect the size or number of sprayers required. For example:
- the number of farm staff employed will reduce if someone retires or leaves and is not replaced
- a neighbouring farm may be taken over, thus increasing the area to be sprayed
- an enlarged cropping area in the future may require a wider grain drill, thus resulting in a change in the tramline width and therefore the sprayer boom width

Existing and future labour and machinery requirements need to be considered as their availability at peak periods may require a change in equipment (e.g. one sprayer instead of two sprayers or a self-propelled sprayer). It is important to look at existing and future tractor size: decisions will be affected by power requirements for carrying or towing a sprayer.

Liquid fertiliser application, via a sprayer that also sprays agricultural chemicals, may be considered now or in the future; such a sprayer can help reduce sprayer costs per hectare.

Timeliness
Timeliness of spraying is very important to the grower; the chemical must be applied at the correct time to ensure its success. The following points will affect timeliness of application:
- area to spray per season
- frequency of spraying
- land characteristics
- weather
- workload of the farm

Area to spray per season
Crops must be sprayed at the correct time. This necessitates a good spraying system (i.e. a sprayer with a good standard of management). The area that a sprayer can cover will depend on a number of factors, one of which is sprayer size. The size of the sprayer can be calculated by the peak spraying requirement, for example, on a cereal farm this peak may occur in April. The Regional Meteorological Office or past records might show that April has x days available.

Example
If the farmer at Acorn Farm has 250ha to spray in April and only ten suitable spraying days, then a spraying system capable of at least 25ha/day is required. When farmers compare various sprayers in action on the farm, they will be able to see if the spraying system attains a minimum of 25ha/day. A higher output will be required depending on the cropping policy, weed/pest/disease infestation and any future changes in farm policy. As timeliness is important, a good rule of thumb is to be able to spray all the cereal crops within a three-day period to ensure correct application at the right growth stage.

Frequency of spraying
Intensive arable farmers may have many chemical applications to make, all of which need to be applied at the right time. The number of applications can be reduced in certain circumstances by using tank mixes, but one must ensure that the chemical mix is compatible. One hears stories of incompatible mixes solidifying, the whole sprayer being totally ruined and then needing correct disposal by a waste contractor.

Land characteristics

Field topography will affect forward speed of the sprayer, thus affecting output. Rough ground will cause excessive boom bounce at high forward speeds resulting in uneven application and boom damage. Undulating land will affect the boom height above the target and visibility for matching bout widths, particularly when using a very wide boom. Soil type and its moisture content will affect access to the field. A fully laden sprayer or water bowser, fitted with incorrect tyres, travelling across a wet and heavy soil, can cause tremendous damage to the soil structure. The speed at which soil dries out after a shower of rain will also affect field accessibility. Low ground pressure vehicles or tyres will help to reduce soil damage and increase the number of spraying days.

Farm layout and road conditions will affect timeliness of spraying. Larger farms and estates obviously have greater distances between fields and need very good spraying logistics. The condition of farm tracks or roads and field access will affect sprayer speed and output per day.

Weather

Available workdays depend upon wind and rainfall. Wind speed and direction and the effect of rainfall on soil type will dictate land access. Rainfall will also affect chemical retention on the crop.

Farmers may consider extending the time available for spraying by applying chemicals at dawn, dusk or at night, when conditions may be more favourable, although they must be sure that the chemicals are suitable for application at this time of day.

The Meteorological Office can help the farmer to make crop spraying decisions which are dependent upon the current weather or past weather records. The Meteorological Office can also provide warnings about plant diseases, soil blows and soil moisture. A recent development has been on-farm weather stations which give farmers immediate access to weather information. Stations can monitor wind speed and direction, air and soil temperature, relative humidity, surface wetness, sunshine and rainfall. The information can be retrieved and stored on a computer, forming a useful record of weather data for management purposes.

Workload

Staff may be busy with other tasks at peak periods, therefore labour availability to aid the spraying system (e.g. transporting water bowsers) may be a problem. Timeliness needs to be considered alongside other farm activities throughout the season – the same principle will apply to the availability of tractors to pull or carry the sprayer or water bowser.

Alternative Spraying Techniques

Farmers need to consider novel sprayer designs such as air-assisted sleeve booms or direct injection units. Each new design needs to be carefully assessed to see if the benefits outweigh the extra costs. With increasing legislation concerning the environmental aspects of pesticide application, techniques which improve deposition, reduce drift and reduce tank rinsate must be considered. The introduction of LERAP, a self-assessment of drift potential, needs to be addressed. Tank rinsing devices and self-emptying boom systems reduce the risk of environmental pollution. Closed transfer devices reduce the risk of operator contamination.

Mounted sprayers

The traditional mounted sprayer, fitted on the rear linkage of the tractor, has the advantage of being a compact unit with a relatively small tank capacity. Tank capacity can be increased by the addition of a front-mounted nurse tank, which also helps with weight distribution. Tank size varies from 220 litres to 2200 litres but the majority are in the 450–1000 litres range.

The disadvantage of mounted sprayers is their high ground pressure on standard tractor tyres but the modern farmer should have different size tyres for varying soil conditions.

Modern demountable skid unit sprayers are covered in the section on self-propelled sprayers below.

Trailed sprayers

The trailed sprayer has a larger tank capacity than the mounted sprayer (1000–3500 litres), resulting in higher workrate as a result of less time being spent travelling to refill. The larger tank and resultant weight may cause problems of soil compaction with standard tyres so the farmer should ensure that alternative tyre sizes are available. The sprayer may not follow the tractor on headlands, so the use of a tracking drawbar may be necessary.

Self-propelled sprayers

Available as either a purpose-built sprayer, as a tractor conversion, or as a demountable unit for certain large tractors. The high output sprayer is suitable for the larger, intensive arable farm or contractor. The large tank (1100–2500 litres) gives high workrates and the ergonomically designed cab good visibility and a comfortable working environment. The high cost of such sprayers is accompanied by a high degree of technology.

Some of the self-propelled machines have demountable tanks to give greater flexibility (e.g. granular fertiliser applicators) and this may help spread the cost.

The farmer should ensure that the engine is powerful enough for the terrain. A fully laden sprayer requires a powerful engine, particularly if the sprayer is hydrostatically driven. If the sprayer is to be used for late application of chemicals, then a high ground clearance is required – a number of self-propelled sprayers achieve this by means of drop axles which give a ground clearance of one metre. Some self-propelled sprayers have four-wheel drive, front and rear steering axles and a choice of wheels for high clearance or low ground pressure.

The modern demountable skid unit sprayer can be fitted to tractors such as the Mercedes Benz Unimog and M.B. Trac. The popularity of the JCB Fastrac has resulted in quite a number of demountable sprayers being used. They have most of the advantages of the self-propelled sprayer but with the added flexibility of being a tractor for farmwork when needed and an agricultural tractor when it comes to selling the machine. Tank sizes vary from 1500–2000 litres.

The tractor/sprayer combination with articulated steering uses a special hitch which joins the tractor (minus its front axle) to a standard sprayer. This type of sprayer follows tramlines very well as the pivot-point is midway between the axles. One of the advantages is that an older tractor may be used as the power unit, thus reducing the cost and allowing the tractor to be used or sold separately.

With all types of sprayer the operator must consider the effects of speed when turning with a full tank of spray on a slope, particularly when high clearance wheels are being used.

Modifying an existing sprayer

An existing sprayer may be modified to increase its output or just to update its components. The sprayer owner could, for example, consider increasing boom width, whether by purchasing or hiring a wider boom, (provided the chassis is strong enough).

Many modern components for updating sprayers can be bought via catalogues or via the Internet and can be supplied by nozzle manufacturers and specialist component manufacturers. These very comprehensive catalogues or web pages are illustrated with excellent diagrams to aid on-farm sprayer modification.

A number of manufacturers offer electronic aids which help monitor the sprayer, self-fill hoses, chemical probes, etc. Modifying an existing sprayer is an excellent way of improving the output of a spraying system in times of financial constraint.

Home construction

If the farmer is mechanically minded or has a competent mechanic and a lot of spare time, he may consider making his own sprayer. Sprayer component catalogues are a most useful source of information to aid the construction of farm sprayers.

The alternative to making a sprayer personally is to commission a sprayer from a manufacturer; a number of manufacturers will construct a sprayer to the client's specification. The cost of a self-propelled sprayer may be reduced by using an existing tractor as the power unit, whether in the traditional configuration or as a tractor/sprayer combination with articulated steering.

Contracting

The sprayer owner should consider the role of the agricultural contractor before purchasing a sprayer. Specialist spraying operations, such as the use of low-ground pressure vehicles or high-ground clearance spraying, may require a considerable financial outlay – farmers should consider employing a contractor for such operations. Alternatively, after purchasing a specialist sprayer, a farmer may have the time to establish a business as a contractor and thus help spread the high costs involved.

Aerial spraying is normally a specialised contracting service and can be financially attractive to some farmers, particularly when late applications are required (e.g. potato blight control).

Purchasing A Sprayer

Construction
Durability is required.

Tank
The tank should be made of non-corrosive materials such as plastic, fibreglass or stainless steel and be adequately supported by the framework. Stainless steel is stronger but heavier. Good tank agitation is very important to ensure that the chemicals are well mixed, so check that the pump is large enough, particularly with large booms or if an extra agitator is fitted. Access for tank filling is most important so check the height and ease of filling. Many modern sprayers are fitted with a self-fill hose for water and an induction bowl for chemical filling. The use of tank rinsing aids (small spinning discs or nozzle heads) fitted in the top of the tank is recommended. They reduce the amount of washing water, reduce the time required to wash out sprayer tanks and eliminate operator contamination.

Pump
The choice is between a roller vane, a diaphragm, a diaphragm/piston or a piston pump. The use of a diaphragm or piston pump, whilst more expensive, means that fewer moving parts come into contact with the solution. The farmer may consider such a positive displacement pump as being the most favourable, particularly where a variable forward speed is required. The pump should have a high capacity, particularly when wide booms are fitted, to ensure a good flow to the nozzles as well as providing good agitation for the tank contents. Centrifugal pumps can give very fast filling rates.

Booms
The booms need to be strong and rigid but not too heavy. Boom suspension is very important if boom bounce and yaw is to be avoided; a stable boom will help towards a more even application. Consider any

weaknesses in boom construction and design; cheaper booms will wear and fall apart more quickly. Recent designs include fibre-glass or nylon lightweight booms and the use of ultra-sonic height detection. Boom folding should also be considered; hydraulic folding, whilst more expensive, saves time in the field and reduces operator contamination, whereas manual folding should be easy enough for one man to operate. Boom folding at $1/2$ or $1/4$ tramline width is useful for an odd-shaped area, as are boom section cut-off valves. The boom should be fitted with a break-back device and a boom-tilting device if sloping land is an important consideration. A twin chemical line facilitates a quick change in application rate for booms without turret-nozzle holders and can be useful, for example, when desiccating oilseed rape, in using one set of nozzles for spraying the upper part of the crop, and the other set for penetrating the crop.

Nozzles
One should consider nozzles made from modern materials which are long lasting, colour coded for easy selection and easily replaced. Bayonet fittings are easier to handle when wearing protective rubber gloves. Quick-change turret nozzles, with two to four different nozzles, reduce the time required when changing application rates – particularly useful for the intensive arable farmer or contractor. Different types of nozzle and the rate of wear of different nozzle materials will affect application rate and spraying costs. Modern anti-drip devices use rubber diaphragms, ensuring a longer life and less maintenance.

Filters
Adequate filtration is important to ensure that the sprayer output is maintained and remains accurate; inadequate filtration results in excessive nozzle wear and nozzle blockages. If the farmer intends to use wettable powders and fine sprays extra in-line filters can be fitted. Filter accessibility for maintenance should be considered.

Pipes and hoses
Check hoses for size. Large bores ensure a good flow and help reduce foaming. Check the strength of the materials used (e.g. check that the pipes don't kink and reduce or prevent the flow).

Framework
The frame needs to be light but strong enough for the treatment it may receive on the farm. The overall strength of the sprayer should be considered. The sprayer should be well made using strong, durable materials but not too heavy. A heavy sprayer with a large tank will cause soil compaction on most soils. The choice of tyres will affect the degree of compaction and one should check that alternative tyre sizes are available with any model. Low-ground pressure tyres are most useful if

spraying in late autumn or early spring; similarly narrow row crop tyres give minimum crop damage and high clearance for later applications. The purchaser should check that the sprayer axle allows the fitting of larger diameter tyres as ground clearance is so important in minimising crop damage. The cost of different tyres must be considered, particularly if they have to be imported.

Controls

It is important to have good access to the controls from the tractor cab, particularly if one is applying toxic sprays. The use of electric or cable controls may need to be considered; they add to the cost but help provide a better and safer environment for the operator, allowing him to concentrate on driving accurately and at the correct speed.

Monitors

Monitors are an important aid to greater accuracy. Monitoring systems can be part of a fully automatic constant spray control, if required. To obtain the best from any monitoring system it is necessary to understand fully how the system works.

Ease of attachment

Trailed sprayers are often easier to attach than mounted sprayers – a lot of time can be wasted with some sprayers if they are difficult to attach. A number of manufacturers use a lower linkage hitch for their mounted sprayers. Other manufacturers mount the pump on the sprayer frame, thus saving time wasted trying to fit a pump and torque chain on to the tractor. A number of manufacturers offer skid units to slide on/off the back of certain tractors such as M.B. Tracs, Chaviots and JCB Fastracs for example.

Cost

The capital cost of a sprayer is very important, as is its resale value so check that the sprayer holds its value. Alternative methods of finance such as leasing may be considered. The cost may be reduced by using the sprayer more often (e.g. for liquid fertiliser application or using low ground pressure tyres to extend its operating season).

Hiring schemes are offered by a number of companies, particularly liquid fertiliser suppliers, and are quite attractive. They often include an annual updating and maintenance service.

Maintenance costs should also be looked at carefully, as these costs can be quite high.

Machinery dealer

Close proximity to a reliable dealer can ensure a speedy service when the sprayer breaks down. The availability and cost of spare parts from the supplier should also be considered.

Ease of maintenance
Good maintenance will aid accuracy and the sprayer should be designed to allow for easy maintenance (e.g. it should be possible to drain it of all liquids to avoid frost damage; filters should be easily dismantled or it should be self-flushing to ensure a good liquid flow).

Operator
The operator should be responsible, well trained and highly motivated. The degree of sophistication of the sprayer may be too great for some people so there is a definite need for operator training. A comprehensive instruction manual should be provided which explains in detail the finer points of the sprayer but training companies such as ATB Landbase and most manufacturers offer appropriate training courses. All operators should attain a level of competency (e.g. National Proficiency Test standard) in order to ensure the safe and correct application of agricultural chemicals. A skilled operator ensures accuracy of application. Many modern sprayers carry an additional water vessel for supplying the operator with clean water for hand washing after making adjustments to the machine. Operator comfort and safety must be given high priority, especially if one is spending many hours spraying during the season.

Dual role
A dual-role sprayer offers the operator a degree of flexibility and in many cases a reduction in machinery costs. Examples of dual-role sprayers include pesticide/liquid fertiliser sprayers.

Personal preference
The final consideration is that of personal preference. This may be based upon one's own experience, gained from many years of crop spraying. Advice may be obtained from a specialist adviser, or neighbouring farmers who have experience of a similar land type and climate. Advice may also come from machinery dealers who, like neighbouring farmers, have experience of local conditions.

After considering all the previous points, one should then draw up a short list of suitable sprayers and then see them demonstrated on your farm, comparing each sprayer under your field conditions and your standard of operation and management.

CROP ESTABLISHMENT EQUIPMENT SELECTION

The cost of establishing a crop needs to be carefully analysed as labour and machinery costs involved can amount to over 50% of the total cost of growing a crop. Seedbed creation can amount to over two thirds of the establishment costs so offer great scope for reduction – some farmers are able to reduce them down to 30%. A cheaper tilth is required. Farmers should consider:

- reducing machinery costs by reducing cultivations without affecting yield
- increasing yield by reducing soil damage through good timeliness
- reducing machinery costs by optimal rather than excessively high power use
- increasing yields via pan busting, if necessary
- selecting the best methods of tilth creation for your farm

Establishment costs have attracted much attention in recent years with the introduction of power harrow/drill combinations, large cultivator/drill combinations, no-till drills and minimum cultivation techniques. Combining secondary cultivation equipment with a drill often reduces the number of tractors and farm workers or, alternatively, allows a much larger area to be drilled. For many years a number of farmers have been guilty of recreational tillage, creating too fine a seedbed with too many passes, resulting in compaction from wheel marks.

High output establishment equipment offers the following advantages, in many cases:
- reduced number of passes
- improved timeliness whilst conditions are good
- reduced power and labour costs/ha

Soil type, weather and power requirements will affect the decision. High output equipment is usually large, heavy and requires a large tractor with good traction.

The output of a large cultivator/drill combinations needs to be calculated using the computer programme Workrate, described in Chapter 4 and Appendix A. Details such as rate of filling, transport time and output in all seasons should be assessed by seeking advice from consultants or visiting farmers who already own such equipment.

Research from Long Ashton Research Station in the Less Intensive Farming and Environment (LIFE) project, provided guidelines for comparative establishment costs and studied energy use for different arable crops and establishment methods. See Donaldson *et al* (1994).

The total cost of large cultivator/drill combinations and tractor needs to be calculated using the machinery cost calculator in Appendix B and described in Chapter 7. Care must be taken if considering a secondhand machine as replacement coulters on such large machines are expensive and this should be reflected in the purchase price.

The costs associated with establishing a crop by traditional means such as plough, discs and cultivators must be weighed against the costs of a modern technique. A partial budget and break-even analysis, described in Chapter 7, should also be carried out as a number of contractors are offering establishment services.

Once the financial implications of a change in establishment techniques have been verified, farmers need to consider the implications for land, labour and capital.

FACTORS AFFECTING THE VALUE OF FARM EQUIPMENT

Soil and terrain
Regional variations in soil mean that different equipment will be required (e.g. heavy clay soils need stronger equipment creating a greater demand for power harrows, semi-digger ploughs etc). Shallow soils with rocks require trip legs on ploughs and cultivators. Many hill farmers often prefer basic, low-specification equipment to cope with the terrain.

Cropping programmes
The range of crops grown within a region (e.g. potatoes, sugar beet, maize, grassland) dictates the need and therefore the value of machinery.

Economics
If a good farming economy exists then there is demand for good quality machinery and prices remain strong. A financial depression in agriculture stifles demand and weakens secondhand values. A strong overseas economy brings about demand for tractors and combines although some imported equipment is sometimes in demand in the country of origin.

Manufacturer
A good, well-known, reputable manufacturer with a good marque will enjoy strong demand and maintain high values. Often marketing programmes, such as subventive financing, subsidised finance offers such as 0% finance schemes, and/or extra discounts on new machines will have a marked effect on secondhand values. Farm machinery from the former Eastern Bloc manufacturers is inexpensive to buy new, causing residual values to be low.

Dealers
The location and reputation of local dealers and their supply of spare parts affects values.

Costs
The cost of repairing older machines becomes too expensive if the price of spare parts rise (e.g. imported machines from a country with a strong currency such as Germany).

The machine
Specialist machines manufactured in low numbers, such as pea viners and bean harvesters, will generally maintain a relatively high residual value but only in the regions where there is demand. Large tractors and

high capacity implements become difficult to sell after two to three years (e.g. a 6m cultivator drill that has sown 2000ha and needs a new set of cultivator tines). A high-capacity, used combine may be attractive to smaller farmers who can buy at a reasonable price and get a better output than if they bought a newer, smaller combine.

Simplicity is often preferable to sophistication but some farmers are able to cope with complicated electronics. On the other hand, complex machinery can be difficult to repair and becomes unpopular in auctions, commanding less money. Machines failing to perform or exhibiting poor reliability soon become known in the farming community and rapidly lose value but the opposite is the case with good machines.

Equipment used for a distinct season, such as hay balers, sprayers and combine harvesters are subject to price fluctuations. Pre-season and early season values are higher as demand and therefore price tails off towards the end of the season. Out-of-season bargains are to be found.

Condition

Appearance is very important. If the exterior is kept intact and free from rust it will be easier to sell, and will maintain value. If all the components are in working order, all levers are still operating and no parts are missing, the good condition will extend the life of the machine and enhance its value. Storing machinery properly protects its appearance and lengthens its life whilst maintaining value. A full service history enhances marketability and subsequent sale.

The market place

Fashion affects the value of equipment (e.g. square ploughs and soil conditioners). When a novel idea is in the ascendancy, values remain firm. Most farmers are fairly astute and wait a few years before purchasing. A good, wet season usually sorts out novel ideas and results in little demand for such equipment.

Disposals of hire fleets can have a marked effect on residual values (e.g. a fleet of lease tractors being sold by a finance house or a bank foreclosing on a large farming company or contractor). The result is dozens of machines hitting the market place simultaneously, supply exceeding demand and a depressed market.

Valuing Machinery

Farmers need to know the value of their equipment. Valuations can be carried out by machinery dealers, engineers or valuers from land agency offices. The following areas should be noted:

Self-propelled machinery

Major areas to observe are: overall appearance/condition, engine, transmission/hydraulics, tyres and wheels, cab.

The following information should be put in the machinery inventory: make, model/specification, registration number, age, hours, size (hp/operating width/capacity), attachments (headers, loaders, weights, LGP tyres etc).

Implements

Major areas to observe are: overall appearance/condition, tyres/wheels, belts/chains

Enter in the inventory: make, model/specification, serial number, age, size (width/capacity/number of rows or furrows, etc), attachments (markers/applicators, monitoring equipment, etc).

Fixed equipment

Major areas to observe are: overall appearance/condition, safety, belts/chains, mobility (labour/transport/recommissioning)

Enter in the inventory: make, model/specification, age, capacity, e.g. tonnes/hour.

Sources of information on the value of equipment may be obtained from local machinery dealers, advertisements in the farming press or on the Internet and from prices obtained at auctions and farm sales.

Appendix C contains information on conducting a detailed inspection on used equipment before purchase.

Further reading

Davies, B, Eagle, D and Finney, B. 1993. *Soil Management*. Farming Press.

Agricultural Tractors 1990. AFRC Engineering Research. Silsoe Research Institute.

How to choose and use combines. Publ. No.88. SAC, Edinburgh.

The utilisation and performance of combine harvesters 1969. Farm mechanisation studies No. 18. Ministry of Agriculture, Fisheries and Food.

McGechan, M.B., Saadoun, T., Glasebey, C.A., Oskoui, K.E. Estimation of combine harvesting work-days from meteorological data. *The Agricultural Engineer* Vol.44 No. 3 1989.

Donaldson, J.V.G, Hutcheon, J.A., Jordan, V.W.L. 1994 Evaluation of energy usage for machinery operations in the development of more environmentally benign farming systems. In: Arable farming under CAP reform, *Aspects of Applied Biology* 40.

Yule, I. Copland, T.A.and O'Callaghan, J.R.O Machinery utilisation on arable farms. *The Agricultural Engineer* Vol. 43 No. 2 1988.

Bailey, J. and Graham, K. How much machinery do you really need? *Farm Management* Vol. 6 No.10 Summer 1988.

Spedding, A., and Kirby, J. *Members survey of combine harvesters*. The Royal Agricultural Society of England. 1994.

Web sites

http://www.agcocorp.com/home.asp
http://www.aea-online.org.uk/directory/AEA.asp?area=Domestic
http://www.allman-sprayers.co.uk/
http://www.casecorp.com/europe/uk/agricultural/newequip/tractors/index.html
http://www.cat.com/
http://www.CLAAS.com/uk/erntetechnik/ernte.htm
http://www.downs.co.uk/
http://www.farmforce.demon.co.uk/
http://www.frazier.co.uk/
http://www.hardixinternational.com/
http://www.horstine-farmery.com/
http://www.ianr.unl.edu/ianr/bse/ttl/
http://www.knight-ltd.co.uk/
http://www.kverneland.com/prod
http://www.krone.de/lm/lm_e.htm
http://www.micron.co.uk/
http://www.mf7200.com/
http://www.newholland.com/
http://www.oecd.org/agr/code/tractors/index.html
http://www.parmiter.co.uk/
http://www.reekie.co.uk/
http://www.samedeutz-fahr.com/gru_ingl/index.htm
http://www.simba.co.uk/
http://www.tractorweb.com/
http://www.unimog.co.uk/

CHAPTER 7

MACHINERY COSTS

The control of machinery costs is a key factor in improving the profitability of a farm. Machinery costs are by far the most complex of all the fixed costs (Figure 7.1) but it is for this reason that they offer the greatest scope in achieving reductions.

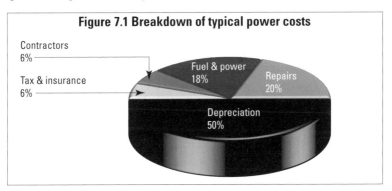

Figure 7.1 Breakdown of typical power costs

Contractors 6%

Tax & insurance 6%

Fuel & power 18%

Repairs 20%

Depreciation 50%

A systematic approach should be used to analyse machinery costs and a good recording scheme should be used to pinpoint where costs are rising. Careful planning will help with future investment decisions and identifying, monitoring and reducing machinery costs will help the farmer face the future with confidence.

It is necessary to know what the costs are because:

● good farm management includes knowing the input costs of the farm business, particularly labour and machinery costs

● alternatives to ownership must always be considered in an attempt to find the least costly method of obtaining farm machinery. Knowing current machinery costs enables comparisons to be made

● new production techniques are always being considered

● future changes in the farm business

● changes in fiscal policy

HOW IMPORTANT ARE MACHINERY COSTS?

On many farms 30–40% of the fixed costs can be allocated to farm machinery. If labour is included, this figure can be as high as 60–65%. On some farms labour and machinery costs may reach a staggering 90%.

Measure machinery costs by keeping good records and comparing with others via farm management data and farmer groups.

FIXED COSTS

Fixed costs are fixed in total and do not vary proportionately with use. They include:
- depreciation
- finance charges
- road fund licence
- insurance
- housing

Depreciation

Depreciation is an estimate of the amount by which the value of an asset falls during the period of ownership. Annual depreciation is a charge against the profits of that particular year which allows the farmer to build up a reserve fund to replace the asset. There are three methods of depreciating farm machinery:

i. Straight line depreciation: original cost minus the sale value divided by the number of years of ownership.

Example
A farm machine costs £20000 and its estimated sale price is £5000 in five years' time

$$£20000 - £5000 \qquad = \qquad \frac{£15000}{5} \qquad = \qquad £3000$$

This is a very simple method of calculating depreciation as it uses an 'estimated' sale price, does not take into account inflation over the 5-year period and the sum set aside each year isn't re-invested to obtain interest.

ii. Diminishing balances: a constant percentage of the written down value. The written down value minus the constant percentage depreciation rate is divided by the number of years.

Example
A farm machine costs £20000 and depreciates at 25% per year

End of year 1: £20000 – 25% (£5000)   = £15000
End of year 2: £15000 – 25% (£3750)   = £11250
End of year 3: £11250 – 25% (£2812)   = £8438
End of year 4: £8438 – 25% (£2109)   = £6329
End of year 5: £6329 – 25% (£1582)   = £4747

This method gives a more realistic, rapid depreciation in the early years but assumes 25% is the correct depreciation rate, that there is no inflation over the five-year period and that the sum set aside each year isn't re-invested to obtain interest.

iii. Sum of the digits: the original cost minus the sale value after the number of years of ownership results in the amount to depreciate.

Example
    A farm machine costs £20000 and its estimated sale price is £5000 in 5 years' time. Depreciation total = £15000
    Sum the digits (add the years to obtain the amount for each year):
5 + 4 + 3 + 2 + 1 = 15
£15000 divided by 15 = £1000 for each year
Depreciation in the 1st year: £1000 x 5   = £5000
Depreciation in the 2nd year: £1000 x 4   = £4000
Depreciation in the 3rd year: £1000 x 3   = £3000
Depreciation in the 4th year: £1000 x 2   = £2000
Depreciation in the 5th year: £1000 x 1   = £1000

This is a more complex method than straight line or diminishing balances.

Choosing the correct depreciation rate is important. Too high a figure may result in paying a balance charge to the Inland Revenue. Depreciation money should be set aside in a replacement fund in order to avoid the problem of an ageing fleet of machinery and no money for replacement. Chapter Twelve details replacement policies to ensure money is available. In periods of high inflation the annual depreciation figure needs to be adjusted to ensure sufficient funds are being put aside.

Choosing a depreciation rate
It is important to note that different makes and models of machine depreciate at different amounts (Table 7.1). There are significant differences in the percentage depreciation between popular Western manufactured tractors and former Eastern Bloc tractors. When considering depreciation rate, farmers need to research the track record of that particular make and model. Information on the secondhand value may be obtained from advertisements in the farming press, the myriad of published auction and 'resale' booklets and from machinery dealers. There are many factors which affect the secondhand value of machinery, as outlined in Chapter Six. Surveys of popular tractors over the past five years will show that not all tractors depreciate at the same amount each year. Depreciation depends upon resale value; trade-in price depends upon 'the deal' being offered at the time of the sale – as such it is little wonder that it is difficult to ascertain depreciation.
    The three depreciation methods discussed above also provide different annual depreciation rates, as shown in the example in Figure 7.2.

Table 7.1 DEPRECIATION METHODS

Depreciation schedule		Based on Average use/annum: 1000hrs	
Average fall in value - % of purchase price			
Frequency of renewal (years)	Complex high depreciation rate e.g potato harvesters, pea viners	Established machines such as tractors, combines, balers etc.	Simple equipment such as ploughs, cultivators, trailers
	%	%	%
1	40	30	20
2	27.5	20	15
3	20*	16	12.5
4	17.5**	14.5	11.5
5	15***	13	10.5
6	13.5	12	9.5
7	12	11	9
8	11	10	8.5
9		9.5	8
10		8.5	7.5

Source: V. Baker, Bristol University, (J.Nix 1999)

* typical frequency of renewal with heavy use
**typical frequency of renewal with average use
***typical frequency of renewal with light use

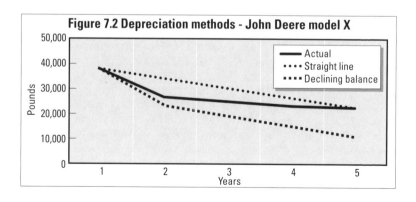

Figure 7.2 Depreciation methods - John Deere model X

Finance charges

The finance charge is the charge on the borrowed capital from the person or company providing the money. If you use your own money it is advisable to charge an 'opportunity cost' as the capital is not available for investment elsewhere.

The finance charge is based upon a rate determined by the financier which may be interest only or interest and capital, depending on the structure and type of loan (Table 7.2). The rate of interest charged depends upon the prevailing bank interest rates at the time of setting up the loan, coupled with many other factors such as risk and security. It is imperative that the prospective borrower understands the basis of the loan, including interest rates and payment dates. It is also important to

Table 7.2 BANK LOAN USING ANNUITY METHOD OF CALCULATING PAYMENTS ON A FIVE-YEAR LOAN							
Year	8%		10%		12%		14%
	I	C	I	C	I	C	I	C
1	80	170	100	164	120	157	140	151
2	66	184	84	180	101	175	119	172
3	52	199	66	198	80	197	92	196
4	36	215	46	218	56	221	67	224
5	19	232	24	240	30	248	36	257

Notes: I = interest payment per £1000
C = repayment of capital per £1000

EXAMPLE
Acorn Farm wishes to borrow £20000 on a five-year loan at 8% interest rate
The loan is made up of interest charges and capital repayments each year

£20000 loan over five years at 8% interest		
	Interest	Capital repayment
Year 1	20 x 80 = 1600	20 x 170 = 3400
Year 2	20 x 66 = 1320	20 x 184 = 3680
Year 3	20 x 52 = 1040	20 x 199 = 3980
Year 4	20 x 36 = 720	20 x 215 = 4300
Year 5	20 x 19 = 380	20 x 232 = 4640

Note: the capital and interest repayment total is approximately £5000 each year
Reducing balance method for calculating loan repayments:
Interest charge for any year = outstanding borrowing multiplied by the interest rate
Capital payment for any year = initial amount borrowed divided by the term

remember that financiers aren't registered charities and are, in fact, extremely profitable businesses.

Interest may be charged on total loan or alternatively, as depreciation allows a proportion of capital to be put aside each year, it may be more realistic to charge interest on average capital employed which is the mid-point between purchase price and sale price.

Interest rates are usually quoted as:

● APR (Annual Percentage Rate). APR is a true compound interest rate and is required by law to be quoted in regulated finance agreements.

● PAF (Per Annum Flat). PAF is a flat interest rate. Interest charge for any year is equal to the initial amount borrowed multiplied by the interest rate and the capital payment for any year is equal to the initial amount borrowed divided by the term.

Example

Acorn Farm wishes to borrow £20000 on a five-year loan at 8% flat interest rate. The loan is made up of interest charges of £1600 per year, every year and five annual capital repayments of £4000 per year. It is interesting to note that at the beginning of the fifth year, only £4000 of the loan remains outstanding. The interest rate in this year is:

$$\frac{1600}{4000} \times 100 \quad = \quad 40\%$$

The amount of the loan outstanding declines over the period as it is paid off. The result is that the true rate of interest (APR) on the loan outstanding is nearly twice the flat rate quoted.

(True rate is the interest as calculated by banks and some finance houses – compounded monthly.)

Road fund licence

The road fund licence is a fiscal charge, levied via government, for tractors.

Insurance

Farm insurance policies can provide comprehensive cover for farm vehicles and implements. They can be based upon a basic cover for a minimum value and then a premium for each £1000 worth of cover up to agreed limits. Sometimes comprehensive insurance will cover damage to attached implements (except ploughs).

Housing

Rain, sleet and snow falling upon farm machinery during the winter is doing more than covering it up, it is also eating away at the investment. Sunlight during the summer months will also affect plastics. Storing equipment indoors will result in:

- greater secondhand value
- fewer repairs
- less downtime

The cost of housing farm machinery is important but often farmers are unable to allocate a cost. Maintenance of farm buildings is expensive and if good records of expenditure are kept then costs can be allocated. The erection of a farm machinery storage building can be an expense which can be allocated to the annual machinery cost.

VARIABLE COSTS

Variable costs vary in total and vary directly with use. They include:
- repairs and maintenance
- fuel and oil
- labour

Repairs And Maintenance

Repairs and maintenance will fluctuate throughout the life of farm machinery. The variations depend on many factors including manufacturer, model, standard of maintenance, operator care and use.

In Figure 7.3 note the peaks and troughs of expenditure which occur during years 5, 7 and 9 due to major repairs to the engine and transmission. Repair costs increase gradually with age but also occur as peaks of expenditure due to the need to replace tyres, overhaul the engine and renew a clutch etc.

There have been a number of surveys conducted on repair and maintenance costs. A number of farm management handbooks also contain relevant information. Great variability occurs between individual models rather than makes.

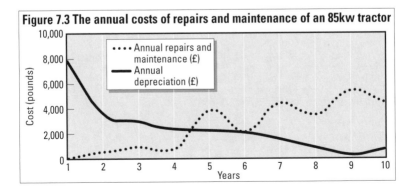

Figure 7.3 The annual costs of repairs and maintenance of an 85kw tractor

Table 7.3 ESTIMATED ANNUAL REPAIR COSTS
(as percentages of purchase price)

	Approximate annual use (hours)				Additional use per 100 hours
	500 (%)	750 (%)	1000 (%)	1500 (%)	ADD (%)
Tractors	5	6.7	8.0	10.5	0.5

	Approximate annual use (hours)				Additional use per 100 hours
	50 (%)	100 (%)	150 (%)	200 (%)	ADD (%)
Harvesting machinery					
Combine harvesters, self-propelled and engine driven	1.5	2.5	3.5	4.5	2.0
Combine harvesters, pto. Driven, metered-chop forage harvesters, pick-up balers, potato harvesters, sugar beet harvesters	3.0	5.0	6.0	7.0	2.0
Other implements and machines					
Group 1 Ploughs, cultivators, toothed harrows, hoes, elevator potato diggers — Normal soils	4.5	8.0	11.0	14.0	6.0
Group 2 Ploughs, cultivators, toothed cutter-windrowers	4.0	7.0	9.5	12.0	5.0
Group 3 Disc harrows, fertiliser distributors, farmyard manure spreaders, combine drills, potato planter wither fertiliser attachment sprayers, hedge-cutting machines	3.0	5.5	7.5	9.5	4.0
Group 4 Swath turners, tedders, side-delivery rakes, unit drills, flail forage harvesters, semi-automatic potato planters and trans-planters, down-the-row thinners	2.5	4.5	6.5	8.5	4.0
Group 5 Corn drills, milking machines, hydraulic loaders, simple potato planting attachments	2.0	4.0	5.5	7.0	3.0
Group 6 Grain dryers, grain cleaners, rolls, hammer mills, feed mixers, threshers	1.5	2.0	2.5	3.0	0.5

When it is known that a high purchase price is due to high quality and durability, or a low price corresponds to a high rate of wear and tear, adjustments to the figures should be made.

Source: Farm Management Pocketbook, John Nix, 27th edition (1996)

Example

A new £40,000 tractor has no repair costs in the first year because it is covered by warranty (providing the tractor isn't damaged). In second and subsequent years the costs in Figure 7.3 could be considered if extended warranty hadn't been purchased. Similarly, the same make and model tractor, if purchased secondhand would incur similar repair and maintenance charges, but based on the original capital cost.

Note that the repairs for soil-engaging implements in Table 7.3 are based on 'normal' soils. The maintenance costs for soil-engaging implements on chalk and flint soils or repair costs on soils with granite boulders can be double or treble the figures quoted. Operator attitude and skill will also affect repair costs.

Other surveys provide the following information:

● a Cotswold survey shows the following repairs and maintenance cost as a percentage of capital cost:

Age (Years)	1	2	3	4	5	6	7	8	9	10
%	3	7	9	11	12	13	14	15	16	17

American surveys observe that repair and maintenance costs are 1.2% per 100 hours of use multiplied by the original capital cost of the tractor and the ASAE yearbook states that tractor repairs and maintenance, based on a life of 10,000 hours equate to

accumulated hours	% of capital cost
2500	9.8
5000	29.7
7500	56.8

It is important to note that the tractor odometer reading may not be a true indication of actual hours worked (see Chapter Eight).

Occasionally, in the popular farming press, there appears an interesting survey of common tractor components, such as radiator, exhaust pipe, fuel pump, fan belt, etc. They make for interesting reading, but, as always, you get what you pay for. If you own a quality tractor and only have to replace a part, albeit quite an expensive part, once every ten years, it is much preferable to replacing the same part frequently, even if less expensive, on rival tractors.

Fuel And Oil

Information is readily available on fuel consumption from handbooks, some manufacturers' brochures and independent testing organisations.

Research by Hunt in the USA suggests that fuel and lubrication costs represent 30–40% of the total costs of operating a tractor. Average fuel consumption can be based upon: 0.243 litres/hr/pto kW. This figure is based upon the average of all independent tractor tests carried out in

Nebraska, USA. The American author, Liljedahl, observed the same figure.

The University of Illinois observed the percentage of time that tractors operate at selected power levels:

Maximum power (%)	Total time (%)
Over 80	16.8
60–80	23.9
40–60	22.6
20–40	17.5
Under 20	19.2

In order to assess fuel use using the above table, farmers need to consider the workload imposed on the tractor and what implements are used. Using a power harrow all day will use a great deal more fuel than pulling a Cambridge roller.

For most farming operations a tractor will work at an average of 55% of its maximum power during the year.

Average fuel consumption (l/hr) = $\dfrac{0.5 \times \text{maximum pto power (kW)}}{\text{KWh/l}}$

Example

If the value of diesel fuel is 2.43kW hours per litre:

Average fuel consumption (l/hr) = $\dfrac{0.5 \times \text{maximum pto power (kW)}}{2.43}$

OECD tractor test results can show a 10–15% difference in fuel consumption figures. In the 37kW (50hp) tractor band there is a staggering maximum difference of 24%. Whilst fuel cost/litre is regarded as quite low compared with petrol, it can add up to quite a large annual bill on many farms. Fuel efficiency must be considered (Table 7.4).

Table 7.4 FUEL CONSUMPTION, PTO AT WORK				
Make & model	Engine power Kw	@standard speed pto		
		rpm	g/kWh	l/h
New Holland 6635	63.0@2500rpm	2200	241	16.7
New Holland 7740	65.8@2050rpm	1900	240	19.1
MF 4255	69.8@2200rpm	1902	242	18.6
Case MX135	99.0@2200rpm	1877	226	27.0
Fendt 515C	110@2300rpm	2070	240	28.7
John Deere 7710	114@2100rpm	1730	227	32.8

Source: OECD data compiled from the DLG test station

Lubrication costs may be 15% of fuel costs. Lubrication oil is quite expensive and data in the handbook of a tractor or combine harvester will show the amount of oil required. For example, the quantity of oil in the rear axle and hydraulic system can be quite large. If farm machinery is serviced at the correct time and oil changes carried out as recommended, then oil cost is an important consideration.

Labour

There are many factors which go to make up the true cost of labour. These include the basic weekly wage, overtime, employers' insurance contributions, holidays and sometimes the value of accommodation.

According to a survey undertaken by *Farm Contractor and Large scale Farmer* it isn't uncommon for 40% of an operator's productive field time to be lost (Figure 7.4). Of course, briefings and maintenance etc are part of the working day. If farm staff are well motivated, secure, remunerated and trained they will look after machinery much better.

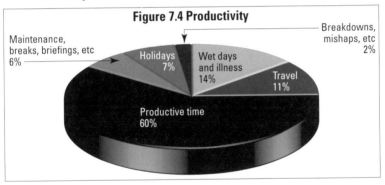

Figure 7.4 Productivity

Maintenance, breaks, briefings, etc 6%

Holidays 7%

Wet days and illness 14%

Travel 11%

Breakdowns, mishaps, etc 2%

Productive time 60%

CALCULATING MACHINERY COSTS

Table 7.5 shows how to calculate machinery costs. A copy of the table, as a computer programme for a spreadsheet, is given in Appendix B. The computer programme can be typed into a standard spreadsheet in ten minutes. By changing hours or hectares worked per year, the user can see the effect of changes in economies of scale etc. Changing any of the inputs can be carried out in seconds. The total cost per hour or per hectare can then be used for comparing costs against machinery rings or contractors' quotations, against machinery hire schemes, etc.

Partial budgets

A partial budget may be constructed to assess the potential financial effect of a partial change in the farm business. In the case of machinery

Table 7.5 CALCULATING MACHINERY COSTS

MACHINERY COSTS CALCULATION

Tractor

PURCHASE PRICE		55000.00	A
SELLING PRICE AFTER	5 years B	30000.00	C
AVERAGE VALUE	A+C/2	42500.00	D
INTEREST	8 % rate x D	3400.00	E
ANNUAL DEPRECIATION	A–C /B	5000.00	F
INSURANCE & ROAD LICENCE	2 % of A	1100.00	G
STORAGE COST	6 % of A	3300.00	H
TOTAL ANNUAL FIXED COST	E+F+G+H	12800.00	
HOURS or HECTARES WORKED ANNUALLY		1300.00	J
FIXED COST PER HOUR or HECTARE	I/J	9.85	K
OPERATING COST PER HOUR or HECTARE			
LABOUR		7.50	L
FUEL		1.90	M
SPARES AND REPAIRS	4 % of A/J	1.69	N
TOTAL COST PER HOUR or HECTARE	K+L++N	20.94	O

management, a partial budget may be used to see the potential effect of a change in policy (e.g. owning a combine harvester versus using a contractor to harvest the cereal crop).

A partial budget considers the financial factors which arise as a direct result of implementing change. These factors include the benefits of change (extra returns and costs saved) and the costs of change (extra costs and returns lost).

Example

Acorn Farm considers using a contractor to harvest 150ha of cereals or to buy a new combine. The computer programme, Machinery Costs (Appendix B), was used to calculate the total annual cost and cost/ha. The following input information was used:

Combine output: 12 tonnes/hr
Average cereal yield: 8 tonnes/ha
Cereal area: 150ha
Purchase price of combine: £120000

Table 7.6 PARTIAL BUDGET

Costs of change	£	Benefits of change	£
Revenue Lost		*Revenue Gained*	
		£10/ha extra yield	1500
New Costs		*Costs Saved*	
Fixed costs			
Annual depreciation	14000	Contractor charges	
Interest 8% on av. capital:	6800	@ £80/ha	12000
Insurance	1800		
Storage costs	1200		
Variable costs			
Repairs/spares/fuel			
@ 32/ha	4800		
Labour @£5/ha	750		
Balance	(-15850)		
	13500		13500

Estimated sale price after 5 years = £50000
Interest rate = 8%
Insurance and road licence = 1.5% of purchase price
Storage cost = 1% of purchase price
Repairs and spares = 4% of purchase price
Labour cost/ha is based upon £7.50/hr, 1.5ha/hr output = $\frac{7.5}{1.5}$ = £5/ha

Fuel cost/ha: £1.90/hr, 1.5ha/hr output = $\frac{1.90}{1.5}$ = £1.27/ha

Contractor's charge = £80/ha
The marginal benefit of timeliness arising from using own combine: £10/ha

From the combine harvester example in Appendix B, it can be seen that the annual cost of ownership amounts to £29500 or £197/ha.

In the example given in Table 7.6 it can be seen that the costs of change amounted to £29350 and the benefits of change to £13500, resulting in a loss of £15850. It is therefore better to use a contractor to harvest the cereals on 150ha.

In order to find out the minimum area at which ownership is a viable proposition, a break-even budget can be created. The break-even point is calculated by dividing the fixed costs of the combine by the variable costs per hectare of ownership.

Fixed costs/year
Old and new variable costs/ha

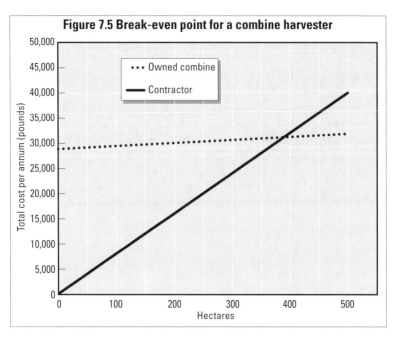

Figure 7.5 Break-even point for a combine harvester

Depreciation + interest + insurance + storage
Contractor's charge – combine running costs

$$\frac{14000 + 6800 + 1800 + 1200}{80 - 18.64} = \frac{23800}{61.36} = 388\text{ha}$$

The minimum area of cereals needed to justify the purchase of a new combine at Acorn Farm is 388ha, quite a substantial area (Figure 7.5). The advantages/disadvantages of machinery ownership compared to alternative forms of acquisition such as contracting are discussed in Chapter 10. The computer programme, Machinery Costs (Appendix B), was used to calculate the total annual cost and cost/ha. The break-even point may also be calculated via this programme. The cost of the contractor or machinery ring may be entered into the spreadsheet cell, D35, and the programme will show two areas of interest:

● total cost/ha, cell D33
● annual savings of using a contractor/machinery ring, cell D38

By changing the area worked annually in cell D21, the total cost/ha appearing in cell D33 will alter. Changing the area worked until cell D33 matches cell D35 will provide you with the break-even area.

It should be noted that many farmers suggest that the above calculations should include the marginal benefit of timeliness arising from using one's own combine, say £10/ha, but this is a debatable point if the contractor arrives at the farm on time.

The financial factors above are just one of the many considerations in the discussion of own machine versus contractors. Timeliness, area to harvest, reliability, availability of labour and personal standards of work must also be considered. Developing a good business relationship with a contractor, planning a work programme months in advance and paying the invoice for the work carried out promptly will help develop a good rapport. Further advantages/disadvantages are discussed in Chapter Ten.

Further reading

Bohm. M. 1993 Breakdowns in agricultural tractors *The Agricultural Engineer* Vol.48. No.3.

Cracknell, J. 1994. Factors influencing the mechanisation of UK agriculture since 1972. *The Agricultural Engineer* Vol.49 No.3.

Donaldson, J.V.G, Hutcheon, J.A., Jordan, V.W.L. 1994 Evaluation of energy usage for machinery operations in the development of more environmentally benign farming systems. In: Arable farming under CAP reform, *Aspects of Applied Biology* 40.

Hunt, D. 1983. *Farm Power and Machinery Management*, 8th ed. Iowa State University Press.

Liljedahl, J.B., Carleton, W.M., Turnquist, P.K. and Smith, D.W. 1979. *Tractors and Their Power Units*. 3rd edition. John Wiley & Sons.

Morris, J. 1988.Tractor repair costs. *Farm Management* Vol.6 No.10 Summer

Turner, M. M. 1996. Depreciation rates of farm machinery. *The Agricultural Engineer* Vol. 48 No.3.

What value your labour? RASE and Deloitte & Touche Agriculture.1996. Stoneleigh: RASE.

Wilson, P. and Davis, S 1998/9. Estimating depreciation in tractors in the UK and implications for farm management decision-making. *Farm Management* Vol. 10 No. 4 Winter.

Yule, I.J. 1995. Calculating tractor operating costs. *Farm Management* Vol 9 No. 3 Autumn.

Web sites

http://www.aadhire.co.uk/
http://www.agmachine.com/
http://www.carteragri.co.uk/
http://www.combineworld.co.uk/
http://www.fwi.co.uk/live/
http://www.nfm.co.uk/index/index.html
http://www.plantmachinery.co.uk/
http://www.peacock.co.uk/
http://www.rossfarm.co.uk/
http://www.watling.co.uk/watland.htm
http://www.reekie.co.uk/used.htm

CHAPTER 8

IDENTIFYING, MONITORING AND REDUCING MACHINERY COSTS

MACHINERY AUDITS, ADVICE AND RECORDS

The aim of any good business is to make sufficient profit to provide reward and cover future investment: cost control is vital. Attention to detail is a key to good management and whilst many farmers will, as a matter of course, control their variable costs through accurate recording, many do not bother to apply the same diligence to the readily discernible elements of their machinery costs.

Machinery Audits And Advice

Farmers need to keep up to date with technical developments, particularly with improvements in machine design and output. Failure to maintain a good machinery stock will lead to an obsolete, ageing fleet of farm equipment. The expense of new equipment must be carefully considered against the cost of good secondhand machines, contractors, machinery rings and hiring. Farmers need to have enough equipment to cultivate and harvest in the work season.

A tour of most farms will show some examples of over-mechanisation. A good machinery audit will concentrate the farmer's mind as to whether a machine is really worth keeping – when was it last used and what is its financial value compared with its agricultural value? Besides releasing some capital, a good farm sale tidies up the farmyard. Before a decision can be taken as to the true value or cost of replacing a machine, the farmer needs to have certain facts to hand and these facts can be obtained via good machinery records and independent advice.

Controlling Machinery Costs

Controlling machinery costs is a key factor in improving the profitability of a farm. Falling margins have resulted in many farmers looking very carefully at their power levels and replacement policies. Surveys of arable and large dairy farms over the past decade have brought machinery management, in particular, the need to consider farm size, soil type and cropping policy with the number of tractors and staff employed, to the attention of many farmers. Typical survey results show economies of scale showing a fall from 2.2kW/ha at 100ha down to 1.3kW/ha at 400ha. It is not uncommon for a farm to have seven tractors but only three drivers, a sign that the farms are over-insured against breakdowns.

Careful planning, using systems analysis, will help with future investment decisions and identifying, monitoring and reducing machinery costs will help the farmer face the future with confidence.

A systematic approach should be taken when considering all farm costs:

- analyse past performance, compare with others
- identify why machinery costs are high
- decide how to reduce machinery costs
- monitor progress

Reasons for high mechanisation costs include:

- too many farm enterprises requiring a large range of equipment with low annual use
- over-mechanisation of many farm tasks which may not result in saving labour
- buying new instead of good quality secondhand equipment
- buying equipment on the misguided assumption that it is tax efficient to do so
- inappropriate financing methods
- replacing equipment in an unplanned method, often prematurely
- excess capacity: too many machines, particularly large machines as a result of over-insurance (i.e. being aware that one year in ten may bring really bad weather which may delay planting or harvesting)
- high repair costs due to failure to follow a regular maintenance schedule, inadequate maintenance, poorly trained operators, and use of local dealers to carry out basic maintenance
- reduced machine output due to the lack of operator skill or motivation
- inappropriate materials handling, either no handlers or very expensive telescopic handlers
- inappropriate use of farm contractors

Reducing depreciation

Obtain a better trade-in for farm machinery via:

- better maintenance
- shelter
- bonus system

Maximise re-sale value via:

- trade-in
- auction
- private sale

Obtain the correct machinery in the first place by:

- seeking independent advice
- avoiding over-mechanisation

Extend planned replacement periods by:
- considering available finances
- considering machine age/hours
- knowing the value of machinery

Consider alternative methods of machinery acquisition:
- short or long-term hire
- machinery rings, contractors or syndication

Provide good operator skills training by:
- encouraging attendance at training workshops
- instilling confidence and pride

Buy quality secondhand machines rather than new because:
- lower purchase price eases cash flow and means lower interest payments
- it may allow equipment update today rather than tomorrow

Reducing repair costs

Select machinery for its simplicity and ease of maintenance:
- is sophisticated equipment really necessary?
- ease of service via central grease points, etc

Record and pinpoint where money is spent on repairs:
- provide straightforward record cards for tasks
- inform operators of the cost of repairs

Provide facilities for better on-farm maintenance and repairs:
- consider the provision of good workshops
- provide a basic tool kit

Choose the best source of skilled labour for repairs
- cheapest is not best
- choose skilled workers with the support of manufacturers

Choose the best source of spare parts:
- *bona fide* parts are superior to cheaper copies
- many components can be sourced directly from manufacturers

Ensure adequate operator training:
- allow operators to go on manufacturer's training courses
- allow time for good preventative maintenance

Develop a bonus system to reward operators who look after equipment:

- consider a profit-sharing scheme to those who reduce repair costs year on year
- consider a prize scheme for those who reduce their 'accident' repair costs

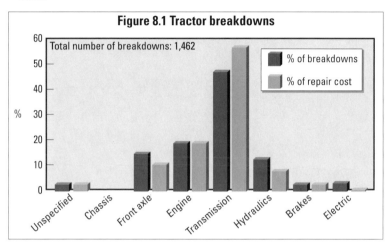

A survey of tractor breakdowns in Sweden (Figure 8.1 and 8.2, Bohm, M. 1993)), shows transmission failure occurs the most (almost 50% of breakdowns), and is also the highest (60%) repair cost.

Figure 8.2 shows how breakdowns were distributed among different machine components in the transmission. The main gearbox, clutch and the speed shift are responsible for over 50% of the breakdowns.

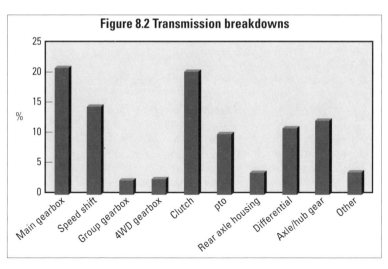

Engineering sophistication to enable simplicity in use, often leads to increased breakdowns and costs.

Reducing fuel costs

Encourage good maintenance:
- allow time for maintenance
- provide facilities for maintenance
- follow maintenance schedules

Test the main tractors with a dynamometer each year:
- winter check-up will show tired engines
- pre-season check will ensure maximum power for demanding operations
- air cleaners must be kept clean

Eliminate unnecessary cultivations and journeys and consider combining operations:
- consider cultivator/drill combinations
- avoid recreational tillage
- do you really need to travel the roads on such large tractors?

Match the correct tractor to the implement:
- know the power requirements for implements
- remove unnecessary weights

Select energy efficient machines:
- consult tractor test results for fuel consumption figures
- compare power requirements (e.g. different designs of forage harvesters)

Adequate operator skills training is necessary in order to ensure fine tuning/adjustment of equipment:
- ensure the correct gears are used
- understand the positions of hydraulic controls

Machinery Records

Dairy farmers keep impressive records concerning the herd, breeding, milk yield and financial data, so why don't farmers apply the same level of record keeping to their machinery stock? Good field records are also becoming a requirement for food production records such as the Assured Combinable Crops Scheme. Today's farmer is faced with many decisions concerning the use of equipment, ranging from a vast array of financial packages to help purchase equipment, to contractors and machinery rings offering services at a known cost and to companies wishing to hire equipment on a short- or long-term basis. Farmers need to know the exact cost of operating equipment on their farms before decisions regarding alternative methods of machinery acquisition can be made. Good machinery records allow farmers to judge if farm equipment is being fully utilised; they help with future investment decisions and aid the construction of a more realistic capital budget. Good machinery records allow management decisions to be based upon facts. Planning will help organise farm resources effectively. The advantage of a flexible plan is that it can be adjusted to changing circumstances.

If machinery recording schemes are simple and easy to use and the purpose and results are explained to farm staff, then staff will accept the extra effort required of them.

Machinery records provide information on:
● the true cost of owning/operating a machine, enabling comparison with other methods of using equipment
● utilisation and performance of the machinery stock
● trends in repairs and maintenance costs, allowing an early identification of problems and indication of the need for remedial action.
● planning of maintenance schedules to minimise 'down time'
● easy access to physical data on each machine as an aid to ordering spare parts
● the need for workshop facilities and equipment

Figure 8.3, which describes the annual costs of repairs and maintenance of a typical 85kW tractor, shows peaks of expenditure occurring during years 5,7 and 9 due to major repairs to the engine and transmission. Repair costs increase gradually with age but also occur as peaks of expenditure with the need to replace tyres, overhaul the engine and renew the clutch, etc.

Sources of information are needed on spare parts and repairs, individual machine fuel/oil record books, labour time sheets, field records, physical and financial records.

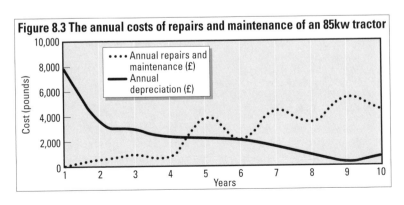

Figure 8.3 The annual costs of repairs and maintenance of an 85kw tractor

Spare parts and repairs

Invoices from parts suppliers offer an extremely useful source of information; besides the cost of the parts they may include the machinery dealer's labour charge. Invoice recording allows the allocation of costs to each machine and helps to provide information on when to replace a machine. Good records will show trends in repairs and detailing individual breakdowns by type may identify rogue machines. Labour hours spent on repairs should also be included.

Individual fuel and oil record books

A simple recording scheme using a book in each tractor or combine harvester will allow the operator to record fuel/oil use. A simple meter fitted to the diesel tank is all that is needed to discover the true fuel cost of a particular field operation. Farmers are able to consider the different fuel consumption rates and costs for various cultivation practices; compare fuel consumption rates of individual tractors and various drivers and compare them to standard data such as the OECD tractor test results. If farmers are aware of excessive fuel use it may enable early diagnosis of engine problems which might otherwise go unnoticed. Early corrective action to fuel injectors/injection pumps may save considerable expense later on.

Labour time sheets

Time sheets inform farmers of the time spent by labour and machinery on a specific task; they also inform farmers as to which task was being carried out when a breakdown occurred and who was driving at the time. A simple,

Table 8.1 ENGINE ODOMETER READING AND HOURS WORKED

Make and model	Engine speed at which odometer is calibrated	Engine speed for 1000rpm pto	Hours shown after 1000hr at 1000 pto rpm
Ford 6610	1666	2060	1236
M-F 690	1500	1900	1266
John Deere	2400	2400	1000
IH 885 XL	1800	2140	1188
Case 1494	2066	2050	992
Fiat 80-90	1800	2380	1322

John Deere has an electronic odometer, which records one hour every 60 minutes regardless of engine rpm.

straightforward time sheet will be filled in correctly by tractor drivers; if time sheets are too complex, the level of 'inventiveness' increases. Good farmers will allocate time for operators to fill in forms correctly. Time sheets also provide a better source of information than the tractor odometer; the odometer only gives hours worked at a certain engine speed or above and this varies from one manufacturer to another (Table 8.1).

Field records

Field records indicate the true machine capacity for your farm with your staff, providing ideal information for labour and machinery planning. Field records also indicate over/under utilisation of equipment and pinpoint investment in too large or too small a machine.

Monitoring machinery records allows farmers to identify small problems which recur on an otherwise good machine; early identification and rectification can prevent an expensive repair bill later on. Identifying such a problem will also reduce the possibility of selling a machine too early; early replacement of machinery before the end of its planned life on a farm will incur a high depreciation.

A growing number of manufacturers are offering information recording systems on tractors, combine harvesters and crop sprayers. Information about machine output and fuel use can be downloaded from in-cab recorders to office computers, allowing faster and easier recording of useful information.

Physical and financial records

Physical details of each machine on the farm can be used to create an inventory of stock. Details on individual machines can include all serial numbers relating to various parts of the machine. One good idea is to carry a postcard of such information in the farm vehicle used to collect spare parts. One is invariably asked for the engine number or filter number or year of manufacture etc, by the storeman at the machinery dealers so a simple record card will be beneficial.

Financial information can include the value of each machine at the year end.

The points below provide the structure for a simple recording system:

- order book is kept in the workshop; order form filled in and passed to the machinery dealer/supplier
- when the part is collected/delivered, the details of the machine are written on the advice note and placed in a lever-arch file in the farm workshop or tractor barn
- at the end of each week the file contents are collected along with time/work sheets and placed in the farm office
- at the end of each month dealer/supplier invoices arrive and are matched with the advice notes
- each machine is coded so the farm secretary/book-keeper can take the advice note and invoice information and place the charge against the respective machine. The work sheet can also be cross-referenced to a machine to check that labour charges and the correct parts have been allocated to the right machine. A running total can be kept of the total cost of repairs and maintenance

● time sheets from the operators show the operations carried out and the time taken. The fuel record book, kept in each machine, can be used to give information regarding fuel consumption

Examples of record cards can be found in Appendix D.

Further reading

Bohm, M. Breakdowns in agricultural tractors. *The Agricultural Engineer* Vol.48. No. 3. 1993
Gladwell, G. Labour and machinery analysis and planning. *Farm Management* Vol. 5 No.9 Spring 1985.
Jarvis, P.J. Controlling fixed costs. *Farm Management* Vol.6 No.78 Winter 1987/8.

Web sites

http://www.aadhire.co.uk/
http://www.agmachine.com/
http://www.carteragri.co.uk/
http://www.combineworld.co.uk/
http://www.fwi.co.uk/live/
http://www.nfm.co.uk/index/index.html
http://www.peacock.co.uk/
http://www.plantmachinery.co.uk/
http://www.reekie.co.uk/used.htm
http://www.rossfarm.co.uk/
http://www.watling.co.uk/watland.htm

CHAPTER 9

FINANCING FARM MACHINERY

Progressive farmers need to be aware of machinery costs in order to remain competitive in the market place. Farmers must also consider methods of financing machinery. There is a myriad of schemes, ranging from outright purchase to hire purchase to leases to various manufacturers' schemes of subsidised finance. The expense of new or secondhand equipment must be carefully considered against the cost of good secondhand equipment, contractors, hiring, syndicates and machinery rings. Good, independent advice must be sought if an expensive mistake is to be avoided.

Understanding Financial Jargon

APR Annual percentage rate. This rate of interest is required by law to be quoted in Regulated Finance Agreements.

PAF Per annum flat. It is the interest calculated as the total repayments less the capital cost divided by the repayment period. The sum is then expressed as a percentage of the capital cost.

True rate Interest as calculated by banks and some finance houses – compounded monthly.

Bank base rate The interest rate at which banks lend to each other, in conjunction with the Bank of England.

Finance house base rate Based on the inter-bank three-month rate and used by the Finance Houses Association.

Variable rate Variable rate finance is quoted as a margin above bank base rate. Interest payments vary according to movements in interest rates.

Advance Amount of money lent after deposit/trade-in.

Payment The amount paid in set periods when you take out a purchase agreement or loan arrangement.

3 + 33 monthly 3-year agreement with 36 monthly payments, 3 paid up front as a deposit.

1 + 11 quarterly 3-year agreement with 12 quarterly payments, 1 paid up front.

1+2 annually 3-year agreement with 3 annual payments, 1 paid up front.

Regulated agreement Agreement covered by the Consumer Credit Agreement.

Unregulated agreement An agreement outside the consumer credit act (e.g. limited companies, local authorities, advances over £15000).

Rental The amount you pay in set periods when you take out a lease agreement.

Primary period The first period of a finance lease when the full capital cost and the interest is paid.

103

Secondary period The period following the primary period.

Peppercorn rental Annual rentals paid during the secondary period until the machine is sold or traded in.

Balloon A final lump sum payment – usually related to expected value of a machine at the end of a lease agreement.

Residual value The value of the machine assumed in calculating the rentals of an operating lease/contract hire.

Interest Rates

Fixed or variable rate

A fixed rate stays at the same level throughout the term of the finance plan. A variable rate changes according to variations in interest rates in the bank base rate. When choosing between fixed and variable rates one should compare the fixed rate with the future volatility of average expected variable rates. Unfortunately it is virtually impossible to know how interest rates will move in the medium to long term. If it seems likely that interest rates will rise and remain high, a fixed rate is probably best. If the farm business is sensitive to changes in interest charges, a fixed rate might be preferable for safety's sake. Otherwise a variable rate is preferable. Overall level of borrowings should also be considered, as should attitude to risk and any cost of 'arranging' a fixed interest loan.

Buy today and use tomorrow's money

The value of money reduces due to inflation and with no inflation real interest rates would be the same as published rates. A 'real' interest rate is one with the effect of inflation removed. As a simple estimate, take the percentage inflation rate from the percentage interest rate. Thus if inflation is 3.5% and interest rates are 8% the real rate of interest is about 4.5%.

Opportunity cost

If the farmer uses his own money to finance a machinery acquisition, what is the opportunity cost?

Example

The best alternative to investing £130000 in a new combine may be to set up a farm shop which will generate £18000 per year.

The opportunity cost of the money invested in the combine is

$$\frac{18000}{130000} \times 100 = 14\%$$

If the farmer purchases a new combine, the investment must generate 14% return on capital through better timeliness and resultant reduced grain loss.

METHODS OF PURCHASING FARM MACHINERY

Outright Purchase Using Your Own Money

Advantages
- capital allowances to set against profits
- there is no interest charge on this money, but it still has a cost – an 'opportunity cost' (the opportunity cost of investing money in a machine is the benefit lost by not investing it in the best alternative)
- takes advantage of 'discount for cash'
- no application to others for money
- capital at the end of a period to put towards another purchase

Disadvantages
- capital tied up in machine
- spending a large sum on one machine can seriously affect cash flow
- capital allowance may be less advantageous than tax allowances for payments on machines leased or hired (100%)
- today's money to pay today's price

Term Loans From A Bank

The loan is repaid by regular instalments of interest and capital over an agreed period.

Advantages
- the farmer owns the asset so may sell it at any time and is able to use sale proceeds towards buying a replacement
- provided the repayments are made, the loan won't be called back in
- the farmer can arrange repayments to match cashflow
- takes advantage of capital allowances
- allows the business to spread the cost of machinery over a period of time for accurate budgeting
- the bank manager is aware of the farm business/financial arrangements
- title to the vehicle or machine passes to the buyer immediately which may be significant where a grant is involved
- buy machinery using tomorrow's money to pay today's price

Disadvantages
- higher rate of interest than an overdraft, especially as overdraft may go up/down as cash flows into the account
- there may be an arrangement fees(1–1.5% of the loan)
- if you are in credit with your current account you are still paying off the loan (compare with an overdraft)
- an initial deposit may be required as security

Bank Overdraft

Bank managers will often agree an overdraft limit with the farmer before the farmer starts operating an overdraft account. Heavy financial penalties will be imposed by the bank if the farmer goes overdrawn without permission.

Advantages
- the farmer owns the asset so may sell it at any time and is able to use sale proceeds towards buying a replacement
- lower interest rate than a loan as the interest calculations are made on a daily basis on the balance outstanding
- takes advantage of capital allowances
- the bank manager is aware of the arrangement
- variable rate of interest
- an overdraft arrangement with a bank allows the farmer to take advantage of 'cash deals' to obtain machinery

Disadvantages
- an overdraft is not a term agreement and must be paid on demand
- uses up lines of credit at the bank which may be better used for short-term funding such as seed or fertiliser purchase
- security is a charge on the business

Hire Purchase

Hire purchase is also known as asset purchase or lease purchase. Cost of equipment plus interest is calculated and divided into a number of payments (usually monthly) over a period of two to five years. The machine belongs to the finance house until the final payment is made.

Advantages
- offered by the machinery dealer through finance house so it can be a convenient, 'one-stop shop'
- payments are fixed and regular
- buy at today's cost with tomorrow's money
- interest rates normally relatively high, but may be discounted by the machine manufacturer/financier
- takes advantage of capital allowances
- potentially dangerous if used to extend credit without bank manager's knowledge

Disadvantages
- equipment is not legally owned by the purchaser until the final payment has been made
- VAT is payable with the deposit. Whilst it may be refunded later, it

can be a large outlay
- generally more expensive and less flexible than leasing
- deposit required, usually in the form of advanced payments
- potentially dangerous if used to extend credit without bank manager's knowledge

Leasing

The machine is bought by the leasing company and rented to the farmer. The leasing company claims initial and capital allowances against tax but the farmer can only claim annual rental and running costs. A third party (e.g. the finance house), sets a future value for the machine based on its expected resale value at the end of the contract. This figure is excluded from the balance on which the rentals are calculated, which means the payments are usually lower than for a loan. The rentals are charged to the profit and loss account as operating costs.

Under a lease, title (legal ownership) will not pass to the lessee (the hirer). Fixed monthly rentals are paid over an agreed period of time. The type and structure of the lease determines the size of payment and the end of contract options. Payments should be matched to the expected period of use for the vehicle in the business. At the end of the rental period the farmer can arrange the sale of the machine on behalf of the finance house and take the proceeds of sale as a rebate of rentals or credit against the rentals of a replacement machine. Alternatively the farmer may extend the rental period for a small annual payment (secondary period).

Types of leases include operating lease (hire or rental scheme) and balloon lease. Operating leases usually contain specific conditions about use and hours/miles per year with penalties if these are exceeded. The advantages and disadvantages of an operating lease are as follows:

Advantages
- rentals are based on the full balance financed (after allowances, part exchange etc)
- rentals are shown as a liability in the balance sheet
- tax relief on rental
- spreads the VAT over the life of the agreement, helping cashflow
- depreciation and rental interest are assigned to the profit and loss account
- hiring the machine – no unknown depreciation or disposal concerns.
- secondary period at nominal cost
- sale at end of primary period may yield cash benefit to go towards next machine
- no initial capital outlay nor capital tied up
- repayments tailored to your cashflow
- no security required as the lease company owns the asset
- allows you to use your own capital for better opportunity costs

Disadvantages
- you don't own the machine
- payments must not exceed the life of the machine
- rentals are at a fixed interest rate
- potentially dangerous if used to extend credit without knowledge of bank manager
- as equipment prices rise, so the replacement cost increases as there isn't a machine to sell or trade in towards the cost of a new one
- the farmer has nothing at the end of the period

A balloon lease is a finance lease in which the payments are structured to include a large one-off payment at the end of the agreed period. This payment is usually the responsibility of the lessee and reflects an estimated disposal value for the machine or vehicle. It is important that the final payment figure is conservative, unless the amount is guaranteed by a third party. The balloon structure reduces the size of the regular rentals throughout the life of the agreement which assists cashflow, and the opportunity to participate in the disposal proceeds can repay careful use. The advantages of a balloon lease are as follows:

- balloon options offer lower overall payments – good for cashflow
- termination option – flexibility to change your vehicle as your requirements change

Contract hire is an operating lease that includes a range of additional services (e.g. servicing, maintenance, road fund licence, etc). Hire rentals are based on depreciation only, not on full capital cost. Contracts are arranged for any number of years and the customer pays rental on agreed dates. The advantages and disadvantages of this system are as follows:

Advantages
- built-in maintenance should ensure the machine operates without breakdowns
- enables farmers to use modern, technically advanced machinery
- allows farmers/contractors to budget more precisely – all costs are known
- allows farmers to release capital to reduce overdraft (e.g. sell off a combine and then hire a combine)
- 100% tax allowance
- no large capital outlay and no deposit helps cashflow
- some companies offer a replacement within 24 hours, particularly combine hirers
- cropping policy can be quite flexible and changes can be made quite quickly

Disadvantages
- as equipment prices rise, so the replacement cost increases as there isn't a machine to sell or trade in towards the cost of a new one
- the farmer has nothing at the end of the period

Securing Finance

Following a review of the machinery policy on the farm and having decided on the manufacturer and size of tractor, a demonstration is arranged with the local dealer. The demonstration tractor is driven by various members of the farm staff with a variety of implements under differing field conditions. Discussion follows as to its suitability. A decision is made and a price discussed and settled upon with the dealer.

The dealer will either arrange finance with a finance company or arrange for a representative of the finance company to call on the farmer. A finance scheme is arranged, the tractor delivered and the farmer pays the finance company according to the finance agreement (Figure 9.1). The dealer receives the price of the tractor and, depending on the arrangement, the finance company owns the tractor until the farmer finishes the agreement, see case studies below.

It is important to note that:
- The finance house representative has to reach pre-arranged sales targets, usually based on the amount financed coupled with interest rates. Interest rates can be as high as 4% over the basket rate so there is room for negotiation with the farmer, depending on the size and term of the loan. If the farmer negotiates a lower rate, then care

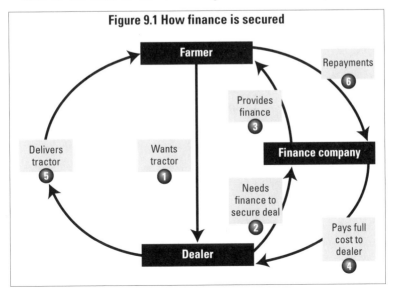

Figure 9.1 How finance is secured

Farmer

Repayments **6**

Provides finance **3**

Delivers tractor **5**

Wants tractor **1**

Finance company

Needs finance to secure deal **2**

Pays full cost to dealer **4**

Dealer

should be taken regarding the amount of deposit, the term of the loan and any early termination penalties. It is important to realise that no-one gives away money for nothing, especially finance houses.

● the financier has to be sure that any monies loaned will be repaid. A number of important factors are considered before an arrangement is made:

i. the character of the business. Is it a safe business, what is the risk? Track record is taken into consideration – is the business on a sound footing?

ii. the abilities of the management to organise and manage the farm. Are financial accounts sound, are the business projections sound? Does the farmer appear to be a good business manager as well as a good stock or crop farmer?

iii. does the business have the means to repay any loans? Is the business doing well enough to meet repayments? Cashflow projections and profit potential are considered

iv. what is the purpose of the loan, is it necessary for the business, will it improve the business and help support the repayments?

v. the amount of money requested. Is it a reasonable request within the constraints of the farm enterprise?

vi. what is the repayment schedule, does it match the cashflows, the seasonality of farming? Would a better repayment schedule help the business? How much is the repayment, can the farmer afford it?

vii. security. Will the financier have to take a risk – if the business collapses, who else has a claim on the land and machinery? How much personal money does the farmer have invested in the business? What other loans exist? Is the farmer a tenant or landowner?

● 0% finance schemes are arranged between the manufacturer of the farm machinery and the finance house. In order to move equipment from the manufacturer through the dealer to the farmer, marketing people often use 0% as a marketing ploy (see case study 5 below). The manufacturer will actually pay the finance charge (interest rate) to move stock. Often market share is an important goal for manufacturers.

Case Studies Of Finance Schemes

Case study 1: standard lease/HP schemes
Deposit/ trade-in = 40%
3 annual payments, one on signing at flat-rate/annum of 1.5% of balance financed or 5 Annual payments at flat-rate/annum of 2.75% of balance.

Example
125hp tractor
Retail   £59186 +VAT
Trade-in deposit (min 40% of retail)   £24000

Balance   £35186
3 annual payments, one on signing, at 1.5% PAF.
£35186 + 1.5% x 3/3 £12256.46
or 5 Annual payments, one on signing at 2.75% PAF of £ 8004.82

Case study 2: contract purchase and rent-a-tractor
The farmer at Acorn Farm is considering obtaining a 125hp tractor.
Invoice price = £51,787.75 + VAT

Contract Purchase over five years
At the end of the finance period an option of:
● Purchase with a 'balloon' payment or
● Re-finance balloon for further period (e.g. over 2 years estimated at two further annual payments)
● Trade-in with minimum guaranteed value above value of balloon in original balloon payment stated above and retain up to 90% of any excess profit if actual trade-in value exceeds minimum

5 annual payments, one on signing of	£10177
and a final balloon payment of	£14025
or	
Guaranteed trade-in	£15980
(n.b. VAT on purchase price with first payment)	
Cost/hour for 1500 hours/year	£6.78

Rent-a-tractor (operating lease)
Can be a total tractor cost package including maintenance
Rentals allow 1500 hours/year usage
At end of hire, tractor is returned or re-hired for a further period

5 annual payments, one on signing, of	£9256
+VAT	
VAT added to each annual rental	
Cost/hour for 1500 hours/year	£6.17

Case study 3: hire purchase or hire scheme (operating lease)
Acorn Farm wishes to obtain a large combine harvester. Should they use hire purchase or a hire scheme (operating lease)?

Retail price:	£173453
Less part exchange and discount	
Amount to be financed:	£86000

1 annual down and 2 annual payments of: £28666.67/year (0%)
or a hire scheme of 200 hours/years:
1 annual down and 2 annual payments: £13425/year plus VAT

At the end of three years the hired combine is returned and the farmer then needs to hire another one. The company policy is to keep you hiring combines as it maintains production at the factory, keeps market share, etc, while the farmer operates the latest combine. The very low hire rate quoted, about £10000 less than other companies, is possible because of the very high resale value due to a strong export market for secondhand combines. An adverse change in exchange rates will lower resale values, stop export trade and increase hire charges.

Case study 4: reducing balance loan, or balloon lease
A Land Rover Discovery is advertised in the farming press for £15740. The vehicle is secondhand, one of many placed into the market at the end of a fleet lease arrangement.

The advertisement states £15740 cash or a personal contract purchase of:
£3756 deposit
plus 36 months of £199
plus a final payment of £8000 = £18920 total

This deal is a lease with a balloon of £8,000 at the end of 36 months. At the end of the lease period the farmer can return the vehicle and have nothing or could consider another finance deal to fund the balloon payment.

Compare with a reducing balance loan:

£3756 deposit leaves £11,984 to finance for 3 years
a. interest charge/year = borrowing outstanding x interest rate
b. capital/year = initial amount borrowed/years of term

Year 1
a. £11984 x 11%   =   £1318
b. £11984 / 3   =   £3394
                         =   £5312 per year or £442 per month
Year 2
a. £11984 – 3394   =   £8590 x 11%   =   £944
b. £11984 / 3   =   £3394
                         =   £4338 per year or £361 per month
Year 3
a. £11984 – £6,788   =   £5196 x 11%   =   £571
b. £11984 / 3   =   £3394
                         =£   3965 per year or £330 per month

Loan total for 3 years = £13,615 plus deposit of £3,756 = £17,371
The deposit is the same as for the balloon lease but for an extra outlay per month. At the end of three years the vehicle is owned and has been bought at a lower price.

Case study 5: 0% interest rate
  An ATV is advertised for £5,395 plus VAT (£944) or
  1 + 11 monthly at 0% finance
  Initial payment: £449.62 plus 11 months of £449.58 = £5,395

  VAT is payable with the first payment: £449.62 plus £944 = £1,393.74

Case study 6: interest rate, PAF
  A mobile grain dryer is for sale at £11000 or
  10% deposit and nothing to pay until the end of harvest (October) next
  year and for the following two Octobers at 1% PAF interest rate.

  10% deposit: £1100, therefore £9900 remaining at 1%

  Interest for any year: initial amount borrowed x interest rate:
  £9900 x 1% = £99

  Capital payment/year: £9900/3 = £3300
  Total/year: £3399

  At the beginning of the last year: £3300 loan remains outstanding, the
  interest rate is:
$$\frac{99}{3300} \quad = \quad 3\% \text{ not the 1\% advertised}$$

Case study 7: Low interest versus a good trade-in or discount
  The local machinery dealer offers Acorn Farm a medium-sized tractor
  for £40000. The manager is offered a choice:
  Scheme 1: £2000 trade-in on the old tractor with credit at 5% flat over
  4 years.
  Scheme 2: £6000 trade-in on the old tractor with credit at 7% flat over
  3 years.
  The manager must first of all calculate the total credit cost.
  The total credit cost of scheme 1: £38000 x 5% x 4 years = £7600
  The total credit cost of scheme 2: £34000 x 7% x 3 years = £7140
  Scheme 1 has the advantage of a lower credit cost per annum, which
might suit the cash flow but Scheme 2 has the dual advantages of a lower
total credit cost and earlier ownership of the tractor.

Cash Or Credit

A large outlay of money, particularly a lump sum, ties up capital,
restricts liquidity and may not make full use of available tax allowances.
A large cash purchase may place an unnecessary burden on the cashflow
of a business. It often makes more sense to spread the cost of new
vehicles or machinery over a period which matches the use you make of

them. Cash is a scarce resource and remains the most liquid funding open to the business. In many businesses there is an over-dependence on short-term sources of finance (e.g. overdrafts), to fund capital investments/purchases. The use of an overdraft to fund a machinery purchase is not advisable as it reduces your lines of credit. Machinery purchases are best made with cash or medium-term finance schemes.

Summary

- match funding to the type of investment (i.e. short, medium or long term)
- consider the financial options, choose the most appropriate/cheapest option
- avoid over-dependence on cash or overdraft as a source of capital
- machinery investment should be planned to match cashflow and budgets. Ensure you will be able to honour your commitment
- compare the total costs of alternative facilities on a like-for-like basis. Do not rely only on interest rates but also consider term, payment schedules, total credit charges and taxation implications, etc
- consider the time value of money
- choosing the most suitable finance option depends on your business circumstances and personal preference
- always take professional advice from specialists, including accountants, regarding taxation before making a decision. Keep advisers informed
- consider alternatives to ownership
- narrow the cost between buying and selling – maintain your equipment

If farmers and growers are considering a finance company, they should select one that provides a range of cost-effective finance options with allowance for seasonal income variations, fixed or variable interest rates and a range of repayment periods and patterns.

Further reading

The cost of money. RASE & Deloitte & Touche Agriculture. Stoneleigh. 1997.

CHAPTER 10

ALTERNATIVE METHODS OF ACQUIRING FARM MACHINERY

Farmers should consider a number of alternatives in order to maintain efficiency at least cost. Contracting work in or out (using excess capacity on neighbouring farms) can help reduce costs. Machinery used infrequently, such as hedgecutters, or machinery which is very expensive, such as precision chop foragers, may be better contracted in.

Machinery rings are an interesting alternative. The idea, from Germany, of matching local farmers' equipment demand with surplus machine capacity appears to have tremendous potential in certain areas.

Hiring machinery enables you to keep abreast of modern equipment without spending a large amount of capital. The hire charge can be offset against tax as a business expense, and so it is worth considering. A growing number of farm machinery manufacturers and dealers are interested in tractor hire schemes. Short-term hire of large tractors is of interest to dairy farmers – for example, to power precision-chop foragers. Large arable farmers could provide their neighbours such a tractor, complete with a driver.

Syndication has always been an area of interest. Group ownership of sugar beet harvesters is popular in certain areas and with good organisation, syndication of other types of machinery could prove useful to some farmers.

As a prelude to studying alternatives to ownership, it is interesting to consider the advantages of ownership.

SOLE OWNERSHIP

- adds value to your business in the form of an asset
- timeliness – you plant/harvest when you wish to, leading to flexibility in the cropping programme and labour/machinery schedules
- equipment is matched to your situation (compared to using contractors or belonging to a syndicate)
- you have the ability to increase income via contracting in work
- maintenance – you know the equipment's likely performance
- no travelling costs to pay
- there are no charges for management supervision

HIRING

A number of farmers are following other industries in the use of machine hire. Farmers are able to hire equipment for a fixed term;

short-term hire runs from one day to one week up to a few months. Long-term hire can exist from one to three years. Advantages include: machinery costs are known at the outset, breakdown repairs are often carried out by the hirer, modern equipment, and it can be a method of releasing capital via the sale of one's own equipment. Hiring offers:

● greater flexibility
● the ability to match machinery to your annual requirements – very important if your area changes or your cropping pattern is affected by the weather or market conditions
● a fully overhauled machine, ready to work and supported by a comprehensive system which includes all repairs, maintenance, insurance, guarantees, and operator instruction/assistance

However, restrictions on hours of use or area usually limit the use of the machine.

CONTRACTING

Traditionally, farmers have used contractors as a way of dealing with excess workloads. The modern, progressive farmer is well advised to use contractors as part of a planned programme of machinery management.

Contractors can help reduce labour and tractor peaks, machinery costs per hectare and often offer a faster service. Disadvantages may include failure to arrive on time and an inferior service, but both problems may be overcome by planning in advance. A recent survey by MAFF and Reading University (Errington and Bennett 1994), based on a detailed study of 539 farms, gives some important new insights into contracting services. Small dairy farmers tend to sell contract services due to excess family labour and the need to supplement farm income, whereas the larger dairy farmer buys in contract services, particularly for peak labour and machinery demands (e.g. silage making). Small cropping farms tend to buy in contract services because their arable area isn't big enough to justify the purchase of large equipment. The larger arable farms tend to make better use of their own machines on their own land. The advantages of contracting are:

● reduction in peaks of labour and tractor demand
● reduction in labour and/or machines, savings in the cost of labour, overtime and bonus payments, and machinery costs
● many contractors are more experienced and have highly trained operators
● the task is often carried out faster and with minimal disruption to other farm operations due to larger equipment and experienced staff
● improvement in the quality of production such as using modern specialist equipment to make silage more quickly, ensuring the cutting of quality grass with a high 'D' value
● temporary increase in the availability of labour and machinery (e.g.

to assist in silage making where there might only be one or two farm staff on a dairy farm)
- machinery supplied is often newer, technically superior, specialised and usually more efficient and reliable
- skill and resources to repair breakdowns quickly
- cost savings may be made
- can be part of a complete service (e.g. drilling, spraying etc)
- no initial capital outlay which might help cashflow
- may allow equipment to be sold to aid cashflow

Objections to contractors are that they don't always come when you want them and that sometimes they do inferior work. Contractors argue that they prefer a planned system compared to a fire brigade approach and that, although there are a few cowboys, there are many established, competent contractors. The potential disadvantages of contractors are:

- reduction in quality of workmanship (e.g. contractors may be in a hurry and may be less careful than own drivers)
- risk of a contractor not arriving exactly when required, with resulting loss of production
- increased management skills needed in some cases (e.g. organisational skills)

Stubble-To-Stubble Contracts

The use of contractors to carry out stubble-to-stubble contracts is increasing within the arable sector. A growing number of farmers are selling their own equipment and relying on contractors to provide labour and equipment although the farmer continues to use management skills for short-term planning, cropping, fertilisers, etc. The major advantages are known labour and machinery costs per hectare (which are often reduced in comparison with the farmer's own costs), flexibility of fixed costs, and, in certain cases, better timeliness. The major question which may be asked is why continue farming if you aren't physically and actively involved? However, the benefits of this approach are:

- it allows land owners to be involved in the management of the farm whilst retaining land and farming interests
- in many cases better timeliness of operations may be obtained by using high capacity equipment
- it allows a flexible farming policy during periods of transition and uncertainty. Allows the individual to move in/out of practical farming
- it allows aspects of the farm business to be budgeted for because costs, particularly machinery costs, are known
- it is a method of providing working capital or reducing debt by selling equipment

MACHINERY SYNDICATES

Machinery syndicates can be divided into three sets: formal groups, informal groups and co-operatives.

Informal Groups

Arrangements are made between two or more farmers who own farm equipment and who co-operate with each other to create a full set of equipment for a given operation (e.g. silage making, mower/conditioner, forage harvester, transport and buckrake). The farmers can cut fixed costs and work as a team to make silage on each other's farms. This form of machinery sharing usually operates on a 'gentleman's agreement' principle without formal rules.

Formal Groups

Syndicates offer farmers less capital outlay, more modern and often larger equipment, with the benefit of less dependence on overtime and casual labour. Disadvantages include finding willing and suitable partners, machine availability and potential lowering of maintenance standards. Two or more farmers jointly purchase an item (or items) of equipment which becomes the property of the syndicate. Syndicates are required to have certain rules with defined areas of responsibility.

The advantages of a machinery syndicate are:
- better use of capital for larger, and technically better equipment
- there is sometimes a higher secondhand value
- it allows labour and equipment from several farms to join forces to create an effective work team
- it allows better use of equipment over larger areas

The disadvantages are:
- a loss of independence
- it requires management agreements (e.g. on who does what, when and to whom)

Sugar beet harvesting is the most popular example of syndication. The advantages here are:

- all management responsibility is taken by the manager
- it provides an opportunity to minimise costs
- it allows harvest permits to be pooled and for example, the heavier land to be harvested first
- it provides an opportunity to increase the workforce temporarily
- it is an opportunity to hire out farm labour to neighbours during slack periods

The disadvantages of a syndicated sugar beet harvesting team include loss of independence and the potential problem of group activities leading to compromise.

Before organising a syndicate a feasibility study should be conducted:
● set out the objectives
● find like-minded farmers in your area who are team players
● analyse the farming businesses
● ascertain the effects of the syndicate on the individual, their income and lifestyle
● examine the legal aspects, such as forming a company
● consider the organisation and management structure of the group. Ensure that one person is in charge, that they have leadership qualities and will maintain high standards of fieldwork and maintenance and motivate team members
● detail the management requirements (e.g. day-to-day supervision, records and payments)
● ascertain the capital required and investigate sources of finance
● detail the labour requirements necessary for the team
● consider machines which are capable of maintaining high output so that no one is disadvantaged by delay

When drawing up a machinery syndicate it is vital that all the members know the rules. Operating rules should include the following.
The individual member should know:
● his proportion of the contribution towards the capital cost of the machine(s)
● the basis for contribution to the running costs (e.g. fuel, labour and transport)
● whether employee insurance covers working away from home

The syndicate as a group should know:
● the entitlement of members to use of the machine(s) and, if appropriate, a rota of use and periods in which each member may retain the machine(s) at any one time should be drawn up
● who will operate, maintain and store the machine
● who will provide labour to the group
● who will provide fuel, tractors and trailers
● how adjustments will be made between members for items provided by one member for another
● who is responsible for insurance and ensuring compliance with safety regulations
● which agent shall be engaged to inspect and report on the machine and when
One member should be appointed as secretary to keep accounts, records, minutes and open a bank account in the name of the syndicate

CO-OPERATIVES

Usually a substantial number of farmers or growers are involved as shareholders in a co-operative. The operation is organised and run by employed staff with farmer representatives sitting on the board. Co-operatives are generally found amongst intensive growers of field-scale vegetables such as potatoes, vegetables and salad crops. The co-operative provides machinery for sowing/planting, harvesting, grading, packing, etc., as well as advice to growers and is responsible for marketing the produce.

MACHINERY RINGS

Machinery rings offer farmers a method of reducing the amount of capital tied up in equipment and the advantages of contracting. A group of farmers who have identified an imbalance in their labour or machinery capacity (this could be an excess or a shortage and both could be on the same farm) then co-operate to utilise the excess capacity on each other's farms.

The ring is administered by a manager and financed by members' subscriptions and a levy on each transaction. The advantage to the supplier of the equipment is that payment is made via direct debit from the demander's bank account within one month. Many contractors have joined rings, ensuring earlier payments than was usually the practice in the past when dealing directly with farmers. One of the major advantages to a demander is that the ring manager can call upon a number of suppliers, thus overcoming the problem of individual contractors not turning up on time.

Benefits of membership include:
● reduction of the fixed costs of machinery ownership via increased annual use and therefore improving utilisation
● contracting work in from another ring member avoids the need to purchase a machine
● contracting in extra capacity supplements existing equipment during periods of need

The disadvantages of membership are similar to those of using a contractor and there is some degree of loss of management over field operations. Also untimely operations may result in lowering of crop yield or quality.

How Machinery Rings Work

Each member of a machinery ring pays a fee related to the size of their business, usually on an area basis. The ring employs an organiser and provides an office with office equipment. The machinery ring does not own any farm equipment. The organiser creates a list of people with machinery available to work for other members (suppliers). The organiser gets a request from a demander and matches up someone to do the work.

A management board decides on a charge for each farm operation. This is reviewed every three months and increased or decreased depending on demand. Contractors play an important part in machinery rings as they increase the size of the machinery pool; they also benefit from having work arranged for them by the ring manager and they are assured of receiving their money within a month (often a cause of concern with direct farm work). A levy of 4% is taken on each transaction, made up by deducting 2% from the sum credited to the supplier and adding 2% to the sum debited from the demander. The demander is given 14 days' notice of the amount to be debited from his account so that he can ensure sufficient funds are available. All financial transactions are by direct debit at the bank.

Further reading

Wright, J. and Bennett, R. *Agricultural contracting in the United Kingdom*. Report No. 21. Dept of Agricultural Economics & Management, University of Reading. 1993.

Errington, A. and Bennett, R. Agricultural contracting in the U.K. *Farm Management* Vol. 8 No.9 Spring 1994.

Evans, M. Methods of machine sharing. *Farm Management* Vol. 6 No.4 Winter 1986.

Evans, M. The benefits and problems of machine sharing. *Farm Management* Vol.6 No.5 Spring 1987.

Web sites

http://www.aadhire.co.uk/
http://www.agmachine.com/
http://www.carteragri.co.uk/
http://www.combineworld.co.uk/
http://www.fwi.co.uk/live/
http://www.nfm.co.uk/index/index.html
http://www.peacock.co.uk/
http://www.plantmachinery.co.uk/
http://www.pyketts.co.uk/
http://www.rossfarm.co.uk/
http://www.watling.co.uk/watland.htm
http://www.reekie.co.uk/used.htm

CHAPTER 11

MAINTENANCE

Maintenance must be carried out if machines are to remain reliable. The ever-increasing cost of spare parts and workshop labour necessitates a good maintenance programme.

Pre-season maintenance is a good opportunity to check over the overall condition of equipment. The earlier it can be done, the longer local dealers have to obtain any spare parts required. Daily maintenance must be carried out according to the operator's manual during the season. Routine maintenance also gives the opportunity to inspect parts for wear and so avoid a costly breakdown. End-of-season maintenance gives the farmer the opportunity to clean equipment down and to inspect parts for wear thoroughly and should be regarded as the beginning of an out-of-season storage policy. The good storage of farm equipment is a must if reliability and secondhand values are to be maintained. The cost of a basic equipment store can be funded by the increased secondhand value realised by well-cared for equipment. A good roof, side walls with plenty of ventilation and a gravel floor is all that is required. A baler with 12 inches of snow on it, quietly rusting away in the corner of a field will prove unreliable in the following season.

Custom, computerised, maintenance management schemes offer the larger farmer many advantages, including:

- planned preventative maintenance
- corrective maintenance
- minor works
- full asset history
- fault analysis
- full machinery reports

Planned maintenance contractors can provide bespoke services for large fleet owners, including routine oil sampling to flag up excessive wear rates, maintenance job cards and maintenance procedures. Oil analysis can detect:

- fuel dilution of lubrication oils
- dirt contamination in lubrication oils
- excessive bearing wear rates
- anti-freeze in the oil

Early detection of faults can:

- increase machinery life
- reduce repair costs, particularly catastrophic breakdowns
- reduce unexpected breakdowns

PRE-SEASON MAINTENANCE: CROP SPRAYER

A safe sprayer, well maintained, will apply pesticides correctly, work without breakdown, minimise waste and be more efficient.

Partly fill the tank with clean water and move the sprayer to an area of waste ground. Remove the nozzles. It is recommended that you wear a coverall, gloves and a face visor when working with the sprayer as it may be contaminated. Engage the drive mechanism and gently turn the pump drive, increasing speed slowly to operating revs. Test the on/off and pressure relief valves, and check the agitation system. Flush through the spray lines, then switch off the engine. Refit the nozzles and check the liquid system again for leaks. The pump, plumbing, booms, nozzles, tank and controls should all be checked for wear and tear. Use a methodical approach, following the flow of water from the tank to the nozzles.

Tank
Ensure the tank is thoroughly clean, with no sediment in the bottom, no fractures and no other damage. Check the agitation is working effectively.

Hoses
Inspect for splits, chafing and cracks, particularly at bends in the line. Inspect connections to ensure they are water-tight.

Pump
Ensure the pump rotates freely and check for leaks on hose connections and on the input shaft. If a pulsation chamber is fitted check the air pressure is at the recommended level.

Filters or strainers
All filter elements and seals must be in place with no blockages. Make sure the correct filter is fitted for the selected nozzles.

Controls
Make sure all electrical connections are clean and free from corrosion.

Pressure gauge
Make sure the needle doesn't fluctuate when the nozzles are spraying and it returns to zero when the sprayer is switched off.

Boom
Inspect the boom for good suspension, break-back and folding. Ensure it moves easily.

Boom piping
Inspect the condition of all hoses/pipes for splits, chafing and cracks,

particularly where the booms fold. Check for leaks around the nozzle bodies.

Check valves
Look out for damaged diaphragms and seats.

Nozzles
Make sure all nozzles on the boom are the same, are in good condition, with no evidence of streaks in the spray pattern. All nozzles must deliver to within + or – 5% of the manufacturer's suggested output.

Controllers
If the sprayer has an automatic controller to monitor the speed of the sprayer and the flow, pressure and area sprayed, ensure it is calibrated and in good condition.

Calibration
Calibration is a vital part of the preparation process for financial, environmental and efficacy reasons. It should be carried out pre-season and in the middle of the season; it is fundamental to calibrate for each intended application rate. Operators simply don't have time not to calibrate.

END-OF-SEASON MAINTENANCE: COMBINES

Make a note of spares required to ensure there will be no delay next season. Beware of rats and mice as they like dark areas with grain so open all elevator/auger trap doors, take out sieves etc. Rust or dirt will affect the operation and wear out belts etc, so take steps to prevent rust. A well-maintained machine has a higher secondhand value.

- clean thoroughly (if using a steam cleaner, beware of over-heating bearings)
- cutterbar knife: remove, oil, cover, lie flat – never hang from each end on a beam
- chains: remove, soak in paraffin oil and store in plastic bags with a label attached. If old fertiliser bags are used, ensure no fertiliser is around to rust the chains. Check sprockets for wear
- slip clutches: check for damage, store with springs slackened off
- table: check damage on tines, auger flights; hydraulic rams should be closed, (if not, grease); remove crop elevator and check
- open all trap doors on elevators/augers
- sieves: clean and remove
- belts: remove or slacken except traction belts
- engine, gear box, final drive: maintenance (e.g. oil change, filters, etc), anti-freeze

- grease all grease nipples
- battery: maintain, monthly trickle charge
- place blocks under the axles and keep the combine off the ground
- bare patches of metal: oil, grease, or paint

FARM WORKSHOPS

The purpose of a well-equipped on-farm workshop is to maximise farm machinery output by minimising the costs associated with breakdowns and lost time. Whether a farm merits the expense of setting up and maintaining a well-equipped workshop, with or without a full-time mechanic, depends on a number of factors. These include:

- farm size, types of enterprise and farm policy
- effects on timeliness
- convenience
- interest and ability of the farm team
- the requirement for routine maintenance and minor repairs
- the need for complex repairs and overhauls
- local machinery dealers, their proximity and workshop costs
- cost of setting up and maintaining a quality workshop

Tidiness is the key to an organised workshop but will the farm team be able to respond to such a request? The good farmer doesn't have time to be untidy – searching around for mislaid items is not acceptable. A good workshop should be well lit, warm in winter and well laid out. The selection of equipment will depend on the level of tasks to be carried out. For routine maintenance by farm staff the expense will be quite small but for major overhauls and a qualified mechanic the costs will be high. Quality tools are best, but expensive, so farmers should acquire them as and when needed in order to build up a collection. Good storage is a must.

Converting existing buildings or setting up a new building as a workshop offers many challenges. A central, well-drained site, surrounded by concrete (to reduce mud) should be considered. High doors to allow the largest machinery inside are needed and the expense of an electrically operated roll-up door needs to be weighed against the extra heating costs incurred by people who are slow in shutting doors. A small personal entry door also helps. One useful idea is to have motion detectors fitted on the walls to switch off lights, etc.

As with most management requirements good record keeping is important. Be aware that workshops come under some very specific Health and Safety legislation so always keep up to date with regulations. Workshops can become 'social clubs' on some farms but good management will ensure they are used for their intended purpose.

Web sites

http://www.accuspray.com
http://www.buyag.com/about.asp
http://www.cals.cornell.edu/dept/aben/pestapp/
http://www.ianr.unl.edu/pubs/NebFacts/nf226.htm
http://www.ianr.unl.edu/pubs/nebfacts/nf225.htm
http://www.ianr.unl.edu/pubs/farmpower/g1261html
http://www.offroadstore.co.uk/parts.asp
http://www.timdixon.bun.com/tim_dixon_macx.html

CHAPTER 12

REPLACING MACHINERY: MAKING THE DECISIONS

The control of machinery costs is very important if the profitability of the farm is to be maintained. Machinery costs comprise depreciation, finance charge, repairs and maintenance, insurance and fuel. Farmers must use machinery depreciation as a fund for replacing equipment if they are to avoid an inefficient, ageing fleet of machines. Farmers must also record increasing repair costs so that they can pinpoint expenditure and replace equipment before an expensive breakdown occurs. There is a need to balance increasing repair costs with decreasing depreciation costs. This chapter reviews some of the methods available to the farmer to plan a replacement policy.

In a survey farmers were asked:

     a. why would you change a machine?

     b. what benefits do you expect from doing so?

     c. when do you think is the right time to make the change?

Typical responses to a. included:

     the machine is worn out

     there is a need for higher output

     the tractor and implements don't match

Typical responses to b. included:

     there is a need for more comfort and less manual labour

     greater reliability is required as there is less time available

     the cost of repairs is increasing all the time

     there is a need to minimise downtime

Typical responses to c. included:

     the machine is obsolete

     it is tax-efficient to make a change

     personal preference

FACTORS TO CONSIDER

● consider whether new machinery is essential

● ascertain if the existing machinery is reliable

● examine the overall farm policy to consider existing and future machinery requirements

● keep detailed records of repairs and maintenance to know the true cost of owning a machine

● discover if discounts or trade-in values exist

● investigate methods of financing new acquisitions

● review workrates and machine capacities to ensure the correct utilisation of existing equipment

- consider the proximity of the agricultural machinery dealer for sales and service
- evaluate existing labour skills against the level of sophistication of the equipment and consider operator training courses
- establish the reputation for reliability of a particular machine
- decide the motivational influence of a new machine on the farm staff
- consider alternatives such as contracting, machinery hire or syndication

Machinery replacement is made easier by planning but factors are variable and each farm will have different criteria governing policy.

The farmer needs to establish the capacity of the business to invest in new machinery – are there annual variations in the business income? If so, does the farmer need to decide on an average amount which can be invested annually?

Knowing when to replace

In principle the number of years a machine should be retained before replacement is the number which minimises its average annual cost, which will certainly fall in the first few years. The annual cost of owning a machine depends on a number of factors:

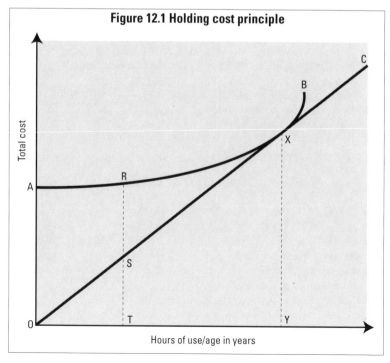

Figure 12.1 Holding cost principle

Total cost

A

R

S

0   T   Y

Hours of use/age in years

● the annual rate of depreciation. Depreciation is the capital cost minus the sale price divided by the number of years of ownership. The rate of depreciation is high in the early years of ownership and declines as the machine becomes older. Depreciation may occur because of obsolescence, gradual deterioration with age, and wear and tear with use

● the annual costs of repairs and maintenance. These are certain to rise as machines grow older, counter-balancing the decreasing cost of depreciation

The combination of the initial capital cost and the cumulative repair costs results in the 'holding cost'. Assuming the repair costs increase with age, the holding cost will take the form of the curve AB (Figure 12.1), OA representing the initial cost. The horizontal axis can represent either hours of work or age in years. Assuming no trade-in value, the

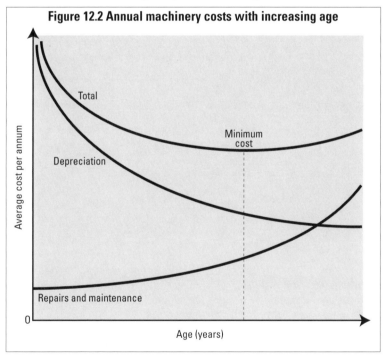

Figure 12.2 Annual machinery costs with increasing age

minimum cost per hour (or per year) will be achieved at OY hours, or years, X being the point of tangency of line OC and the holding cost curve AB, when the cost per hour (or year) will be XY.

Figure 12.2 shows how annual machinery costs increase with age. As depreciation decreases with age, so repair bills rise. Depreciation costs are added to the repairs and maintenance costs to give the total annual costs which reduce initially and rise over time.

It is important to note that the repair cost curve is very smooth in Figure 12.2 because it is shown as an average. Repair bills occur in peaks, due to intermittent levels of expenditure (e.g. new tyres, engine overhaul, replacement clutch, transmission failure, etc).

Example

A farmer wants to decide when to replace an 89kW four-wheel drive tractor which costs £30,000. Repairs, maintenance and depreciation costs are shown in Table 12.1.

The repair costs show the high peaks of expenditure which occur during years five, seven and nine due to major repairs to the engine and transmission. The repair costs increase with age but are in peaks of expenditure.

Table 12.1 THE ANNUAL COSTS OF REPAIRS/MAINTENANCE AND DEPRECIATION OF AN 89KW TRACTOR		
Year	Annual repairs and maintenance £	Annual depreciation £
1	0	7800
2	600	3600
3	1000	3000
4	900	2400
5	3900	2300
6	2100	2100
7	4500	1500
8	3500	900
9	5500	300
10	4500	700

In Figure 12.3, which relates to the annual costs of repairs and maintenance, depreciation is steepest in the first year, and then reduces considerably with time. The tractor has a base value of £5000 and depreciates very slowly between years seven and ten. When the tractor is ten years old it passes a milestone and continues to reduce in value.

The costs in Table 12.1 can be tabulated to show the cumulative costs of repairs/maintenance and depreciation as the average cost as shown in Table 12.2.

The method of deciding when to replace a machine shown in Table 12.2 is very useful if one knows all the information in advance – the use of detailed records will help pinpoint repair bills. The system, whilst somewhat theoretical, can be used on a year-to-year basis (i.e. financial information is recorded and the farmer observes the upward trend in repairs against the downward trend in depreciation).

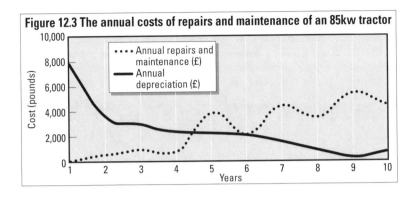

Figure 12.3 The annual costs of repairs and maintenance of an 85kw tractor

Farmers need to know when to replace individual machines but it is also important to ensure that machines are replaced in some order or planned sequence. The farmer can build up a planning chart indicating when to replace equipment over a given time period.

Year (n) (C)	Cumulative repair cost	Cumulative depreciation	Cumulative total cost	Average annual cost (C/n)
	£	£	£	£
1	0	7800	7800	7800
2	600	11400	12000	6000
3	1600	14400	16000	5333
4	2500	16800	19300	4825
5	6400	19100	25500	5100
6	8500	21600	30100	5017
7	13000	23100	36100	5157
8	16500	24000	40500	5062
9	22000	24300	46300	5141
10	26500	25000	51500	5150

Table 12.2 THE AVERAGE ANNUAL COST OF A 9KW TRACTOR

The average annual cost is least in year four, so this is the year the tractor should be replaced.

REPLACEMENT POLICIES

A Basic Policy

A basic machinery replacement policy can be drawn up by creating an inventory of the machines on a farm and then deciding upon the 'life expectancy' of each machine on the farm under its conditions. A chart can then be constructed by listing the equipment against a column of purchase date and life expectancy. The chart, whilst simple to construct, takes no account of peaks in expenditure in years to come, nor does it take into account any money set aside each year as depreciation. This basic method is a good aide-memoire and shows the need for careful planning. The chart is based entirely upon the age of the equipment.

TABLE 12.3. A BASIC MACHINERY REPLACEMENT POLICY

Machine	Year Purchased	Life yrs.	00	01	02	03	04	05
Tractor 134kW	1998	3		X			X	
Tractor 67kW	1999	4				X		
Tractor 67kW	1999	4				X		
Tractor 67kW	1998	4			X			
R T F L T	1994	8			X			
Combine harvester	1995	5	X					X
Baler	1994	8			X			
Bale accumulator	1994	8			X			
Bale loader	1994	8			X			
Fert. spreader	1997	4		X				
Drill, 4 m	1995	8				X		
Sprayer, trailed	1996	8					X	
Trailer 1,	1990	10	X					
Trailer 2,	1990	10	X					
Plough, 6f	1993	10				X		
Subsoiler	1993	12						X
Discs, 6m	1992	12					X	
Cultivator	1992	8	X					
Harrows, chain	1995	12						
Rolls, Cambridge	1989	12		X				

The Whole Stock Replacement Policy

The whole stock method was devised by Mike Barrett, an ADAS adviser, with the intention of arriving at a replacement re-investment strategy which is uniform, has no big fluctuations in demand with all the machines being rotated at reasonable intervals (Barrett, 1984).

The farmer creates an inventory of all the equipment with their proposed replacement intervals. For example, using a 12-year cycle, some machines (e.g. combines) would be replaced four times, some twice and others just once (e.g. trailers). A draft replacement policy is drawn up and a level of re-investment decided upon. It becomes easier to make variations because the effect of postponing or advancing a replacement can be seen. The method can be reviewed every few years.

An example is outlined to show the usefulness of this method when applied to a 300-hectare, completely combinable crop farm. There is one combine which is replaced at three-year intervals, one 134kW tractor (three years) and three 67kW tractors (six years) together with the appropriate tillage and other equipment making an inventory of 22 items (Table 12.3).

Figure 12.4 shows the whole stock method in the form of a bar chart. Note that the replacement cost forms the vertical axis and that the numbers on each bar represent the machines itemised in Table 12.4.

The average annual replacement cost is £42000. Note that inflation isn't taken into consideration. If inflation is running at 5% per annum

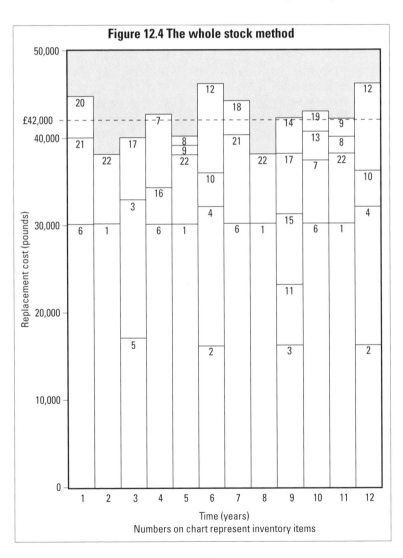

then the average annual replacement cost will have risen to £75425 by year 12.

TABLE 12.4. A WHOLE STOCK REPLACEMENT POLICY

Item No.	Description	Current list price (£)
1	Tractor 134kW	55000
2	Tractor 67kW	30000
3	Tractor 67kW	30000
4	Tractor 67kW	30000
5	R T F L T	25000
6	Combine harvester	65000
7	Baler	10000
8	Bale accumulator	3000
9	Bale loader	3000
10	Fertiliser spreader	5000
11	Drill, 4m grain only	8000
12	Sprayer, trailed	13000
13	Trailer 1, 8 tonne	5000
14	Trailer 2, 8 tonne	5000
15	Plough, 6f reversible	10000
16	Subsoiler	6000
17	Discs, 6m	8000
18	Cultivator, coil tine	5000
19	Harrows, chain	2000
20	Rolls, Cambridge	6000
21	Land Rover	15000
22	Car	18000

The Capital Budget

Planning when to replace and knowing how much to set aside can be made easier by using a capital budget. The budget allows the farmer to maintain a level of investment in farm machinery – particularly useful if the farmer wishes to keep his machinery up to date and remain efficient. In the UK, farmers are allowed to offset part of their machinery depreciation against tax so the wise farmer needs to use this tax allowance to purchase farm machines to keep his fleet up to date. Farmers have often used the depreciation allowance for other purposes, resulting in an inefficient, ageing fleet of machines.

Machinery costs can account for 30-40% of the fixed costs of owning a farm, when combined with labour costs this can amount to 60%. Farm management surveys show that different types and sizes of farm have different levels of investment in farm machinery. A 200-hectare, mainly combinable crop farm in England, invests about £475 per hectare in farm machinery.

Machine	Year Purchased	Life yrs.	'00	'01	'02	'03	'04	'05
Table 12.5 EXAMPLE CAPITAL BUDGET								
Tractor 134kW	1995	5	X					X
Tractor 67kW	1996	5		X				
Tractor 67kW	1999	8						
Combine harvester	1997	5			X			
Baler	1994	10					X	
Fert. Spreader	1997	4		X				
Sprayer, trailed	1998	8					X	
Plough, 6f	1993	10				X		
Discs, 6m	1994	12						
Cultivator	1992	8	X					
Harrows, chain	1995	12						
Rolls, Cambridge	1989	12		X				
Total new cost	£000s		45	41	65	10	23	60
Trade-in price	£000s		30	14	30	2	2	30
			1	4			3	4
				5				
Net change over	£000s		14	18	30	8	18	26

Annual average investment:
£114,000 divided by 6 years = £19,000 per annum.

The rates of depreciation vary for each type of machine (e.g. tractors and combines may depreciate at 15% per year on average, whereas cultivation equipment and farm cars may depreciate at 25%). There are many factors such as make, model, condition, season and markets which affect depreciation rates.

For the following example 18% is used as a weighted average rate of depreciation.

Example

If the average investment in machinery on a 230ha cereal farm is £475 per hectare, then 18% depreciation is £85.50 per hectare.

If the rate of inflation is, say, 5% per year, to maintain a level of investment in machinery, the farmer must invest:

£85.50 + £4.25 = £89.95 per hectare

or

£89.95 multiplied by 230 hectares = £20,688

Table 12.5 shows an example of a capital budget for a farm. The budget takes into account a planned replacement policy against a sum of money put aside each year and accounts for inflation. The example shows that the farmer is nearly investing enough money each year to maintain up-to-date machinery.

Summary

● Farmers need to plan their machinery replacement if they are to maintain up-to-date equipment

● a machinery replacement policy needs to be based upon an inventory of equipment against time

● monies set aside for depreciation need to be re-invested to maintain a level of modern equipment

● a machinery replacement policy needs to be flexible. If there is a good farm profit, the planned replacement can be brought forwards. Conversely, if there is a loss, then the policy can be extended

Further reading

Barnard, C.S. and Nix, J.S. *Farm Planning and Control*. Cambridge University 2nd edition 1979.

Barrett, M.1984. *Machinery replacement policy*. Oxford Bulletin ADAS.

APPENDIX A: **WORKRATE**

The computer programme Workrate has been written to demonstrate how various inputs affect the output of farm machinery. Farmers may use the programme to demonstrate the effects of:

- changing working width (e.g. tramline systems, boom width, drill width etc)
- changing the rate of filling a machine (e.g. large water bowsers, big bags etc)
- comparing the output of different machines to provide comparisons of outputs

A basic spreadsheet programme such as Excel may be used. Programme users will need to ascertain the necessary programming system for their spreadsheet, but the example quoted can easily be inserted into a spreadsheet. The formula used is based upon the workrate formula discussed in Chapter 4.

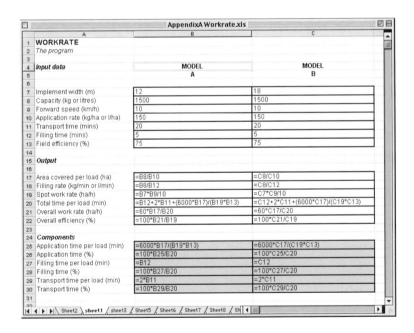

	A	B	C
		AppendixA Workrate.xls	
1	WORKRATE		
2	*The program*		
3			
4	*Input data*	MODEL	MODEL
5		A	B
6			
7	Implement width (m)	12	18
8	Capacity (kg or litres)	1500	1500
9	Forward speed (km/h)	10	10
10	Application rate (kg/ha or l/ha)	150	150
11	Transport time (mins)	20	20
12	Filling time (mins)	5	5
13	Field efficiency (%)	75	75
14			
15	*Output*		
16			
17	Area covered per load (ha)	=B8/B10	=C8/C10
18	Filling rate (kg/min or l/min)	=B8/B12	=C8/C12
19	Spot work rate (ha/h)	=B7*B9/10	=C7*C9/10
20	Total time per load (min)	=B12+2*B11+(6000*B17)/(B19*B13)	=C12+2*C11+(6000*C17)/(C19*C13)
21	Overall work rate (ha/h)	=60*B17/B20	=60*C17/C20
22	Overall efficiency (%)	=100*B21/B19	=100*C21/C19
23			
24	*Components*		
25	Application time per load (min)	=6000*B17/(B19*B13)	=6000*C17/(C19*C13)
26	Application time (%)	=100*B25/B20	=100*C25/C20
27	Filling time per load (min)	=B12	=C12
28	Filling time (%)	=100*B27/B20	=100*C27/C20
29	Transport time per load (min)	=2*B11	=2*C11
30	Transport time (%)	=100*B29/B20	=100*C29/C20
31			

Sheet2 \ sheet1 / sheet3 / Sheet5 / Sheet6 / Sheet7 / Sheet8 / Sh

A. Workrate. The example, quoted in Excel, uses an = sign to denote a formula. Each cell is represented under typical spreadsheet format, e.g. A, B, C, D etc, 1, 2, 3, 4, etc.
Creating the programme – once column B has been inserted – may be speeded up by copying column B, and then pasting it under column C.
B. Workrate – a two-model example, shows the effect of rapid filling on the headland. This reduces transport and filling time. This programme is based upon the basic programme.
C. Workrate – a six-model example, shows how many variables may be considered side by side. The example shown is for a crop sprayer with varying boom widths, tank sizes, rates of fill and efficiencies, (as discussed in Chapter 4). The spreadsheet was created, as in Section A above, but the extra models were created by copying column B, and then pasting it under columns C–G.

	A	B	C	D	E
	AppendixA Workrate.xls				
1	**WORKRATE**				
2	*A two model example*				
3					
4	*Input data*	MODEL	MODEL		
5		A	B		
6					
7	Implement width (m)	12	12		
8	Capacity (kg or litres)	1500	1500		
9	Forward speed (km/h)	9.7	9.7		
10	Application rate (kg/ha or l/ha)	200	200		
11	Transport time,one way, (mins)	5	0		
12	Filling time (mins)	20	5		
13	Field efficiency (%)	75	75		
14					
15	*Output*				
16					
17	Area covered per load (ha)	7.5	7.5		
18	Filling rate (kg/min or l/min)	75	300		
19	Spot work rate (ha/h)	11.64	11.64		
20	Total time per load (min)	81.5	56.5		
21	Overall work rate (ha/h)	5.5	8.0		
22	Overall efficiency (%)	47.4	68.4		
23					
24	*Components*				
25	Application time per load (min)	51.5	51.5		
26	Application time (%)	63.2	91.2		
27	Filling time per load (min)	20.0	5.0		
28	Filling time (%)	24.5	8.8		
29	Transport time per load (min)	10.0	0.0		
30	Transport time (%)	12.3	0.0		
31					
32					
33					
34					

Sheet2 / sheet1 / sheet3 / Sheet5 / Sheet6

```
┌─────────────────────────── AppendixA Workrate.xls ───────────────────────┐
```

	A	B	C	D	E	F	G	H	I	J
1	**WORKRATE**									
2	*A six model example*									
3										
4	*Input data*	MODEL	MODEL	MODEL	MODEL	MODEL	MODEL			
5		A	B	C	D	E	F			
6										
7	Implement width (m)	12	12	18	24	12	24			
8	Capacity (kg or litres)	1500	3000	1500	1500	1500	3000			
9	Forward speed (km/h)	9.7	9.7	9.7	9.7	9.7	9.7			
10	Application rate (kg/ha or l/ha)	200	200	200	200	200	200			
11	Transport time (mins)	5	5	5	5	1	1			
12	Filling time (mins)	25	25	25	25	10	10			
13	Field efficiency (%)	75	75	75	75	80	80			
14										
15	*Output*									
16										
17	Area covered per load (ha)	7.5	15	7.5	7.5	7.5	15			
18	Filling rate (kg/min or l/min)	60	120	60	60	150	300			
19	Spot work rate (ha/h)	11.64	11.64	17.46	23.28	11.64	23.28			
20	Total time per load (min)	86.5	138.1	69.4	60.8	60.3	60.3			
21	Overall work rate (ha/h)	5.2	6.5	6.5	7.4	7.5	14.9			
22	Overall efficiency (%)	44.7	56.0	37.2	31.8	64.1	64.1			
23										
24	*Components*									
25	Application time per load (min)	51.5	103.1	34.4	25.8	48.3	48.3			
26	Application time (%)	59.6	74.7	49.5	42.4	80.1	80.1			
27	Filling time per load (min)	25.0	25.0	25.0	25.0	10.0	10.0			
28	Filling time (%)	28.9	18.1	36.0	41.1	16.6	16.6			
29	Transport time per load (min)	10.0	10.0	10.0	10.0	2.0	2.0			
30	Transport time (%)	11.6	7.2	14.4	16.5	3.3	3.3			
31										

```
│◄ ◄ ► ►│\ Sheet2 / sheet1 \sheet3 / Sheet5 / Sheet6 / Sheet7 / Sheet8 / Sh │◄│
```

APPENDIX B: MACHINERY COSTS CALCULATION

The computer programme 'Machinery costs calculation' has been written to demonstrate how various inputs affect the costs of farm machinery. Farmers may use the programme to demonstrate the effects of:

● changing the fixed costs of ownership such as the purchase price, resale value, finance charge, insurance and storage costs

● changing the variable costs such as fuel and lubrication, repairs, maintenance and labour

● comparing the cost of owning a machine versus selecting a contractor or machinery ring

● ascertaining the break-even point of owning against contracting machinery

A. Machinery costs calculation – the basic programme, shows the formulae used for the spreadsheet. A basic spreadsheet programme such as Excel may be used. Programme users will need to ascertain the necessary programming system for their spreadsheet, but the example quoted can easily be inserted into a spreadsheet. The formula used is based upon the machinery costs formula discussed in Chapter Seven.

The example, quoted in Excel, uses an = sign to denote a formula. Each

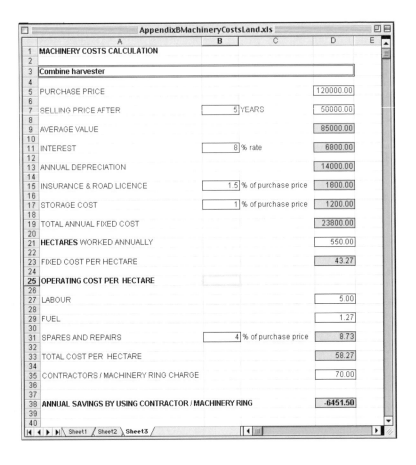

	A	B	C	D	E
1	MACHINERY COSTS CALCULATION				
2					
3	Combine harvester				
4					
5	PURCHASE PRICE			120000.00	
6					
7	SELLING PRICE AFTER	5	YEARS	50000.00	
8					
9	AVERAGE VALUE			85000.00	
10					
11	INTEREST	8	% rate	6800.00	
12					
13	ANNUAL DEPRECIATION			14000.00	
14					
15	INSURANCE & ROAD LICENCE	1.5	% of purchase price	1800.00	
16					
17	STORAGE COST	1	% of purchase price	1200.00	
18					
19	TOTAL ANNUAL FIXED COST			23800.00	
20					
21	HECTARES WORKED ANNUALLY			550.00	
22					
23	FIXED COST PER HECTARE			43.27	
24					
25	OPERATING COST PER HECTARE				
26					
27	LABOUR			5.00	
28					
29	FUEL			1.27	
30					
31	SPARES AND REPAIRS	4	% of purchase price	8.73	
32					
33	TOTAL COST PER HECTARE			58.27	
34					
35	CONTRACTORS / MACHINERY RING CHARGE			70.00	
36					
37					
38	ANNUAL SAVINGS BY USING CONTRACTOR / MACHINERY RING			-6451.50	
39					
40					

Sheet1 / Sheet2 \ Sheet3 /

cell is represented under typical spreadsheet format, e.g. A, B, C, D, etc, 1, 2, 3, 4, etc.

B. Machinery costs calculation – tractor, provides a worksheet for changing the costs associated with owning a tractor. This programme is based upon the basic programme.

C. Machinery costs calculation – Combine harvester, provides a worksheet for changing the costs associated with owning a combine harvester. This programme is based upon the basic programme.

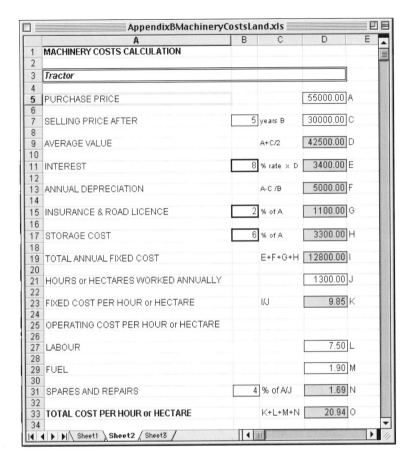

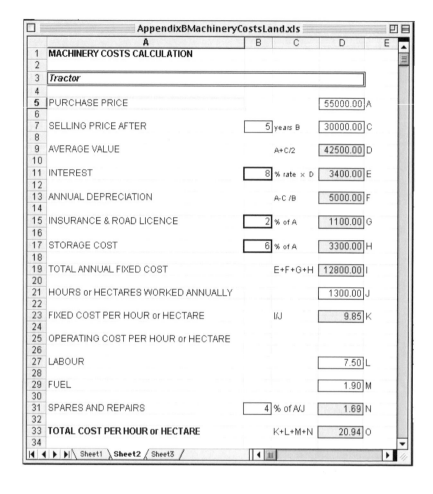

	A	B	C	D	E
	AppendixBMachineryCostsLand.xls				
1	**MACHINERY COSTS CALCULATION**				
2					
3	*Tractor*				
4					
5	PURCHASE PRICE			55000.00	A
6					
7	SELLING PRICE AFTER	5	years B	30000.00	C
8					
9	AVERAGE VALUE		A+C/2	42500.00	D
10					
11	INTEREST	8	% rate × D	3400.00	E
12					
13	ANNUAL DEPRECIATION		A-C /B	5000.00	F
14					
15	INSURANCE & ROAD LICENCE	2	% of A	1100.00	G
16					
17	STORAGE COST	6	% of A	3300.00	H
18					
19	TOTAL ANNUAL FIXED COST		E+F+G+H	12800.00	I
20					
21	HOURS or HECTARES WORKED ANNUALLY			1300.00	J
22					
23	FIXED COST PER HOUR or HECTARE		I/J	9.85	K
24					
25	OPERATING COST PER HOUR or HECTARE				
26					
27	LABOUR			7.50	L
28					
29	FUEL			1.90	M
30					
31	SPARES AND REPAIRS	4	% of A/J	1.69	N
32					
33	**TOTAL COST PER HOUR or HECTARE**		K+L+M+N	20.94	O
34					

Sheet1 \ **Sheet2** / Sheet3 /

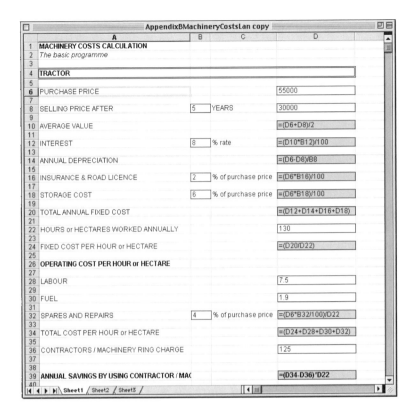

	A	B	C	D
1	MACHINERY COSTS CALCULATION			
2	*The basic programme*			
3				
4	TRACTOR			
5				
6	PURCHASE PRICE			55000
7				
8	SELLING PRICE AFTER	5	YEARS	30000
9				
10	AVERAGE VALUE			=(D6+D8)/2
11				
12	INTEREST	8	% rate	=(D10*B12)/100
13				
14	ANNUAL DEPRECIATION			=(D6-D8)/B8
15				
16	INSURANCE & ROAD LICENCE	2	% of purchase price	=(D6*B16)/100
17				
18	STORAGE COST	6	% of purchase price	=(D6*B18)/100
19				
20	TOTAL ANNUAL FIXED COST			=(D12+D14+D16+D18)
21				
22	HOURS or HECTARES WORKED ANNUALLY			130
23				
24	FIXED COST PER HOUR or HECTARE			=(D20/D22)
25				
26	OPERATING COST PER HOUR or HECTARE			
27				
28	LABOUR			7.5
29				
30	FUEL			1.9
31				
32	SPARES AND REPAIRS	4	% of purchase price	=(D6*B32/100)/D22
33				
34	TOTAL COST PER HOUR or HECTARE			=(D24+D28+D30+D32)
35				
36	CONTRACTORS / MACHINERY RING CHARGE			125
37				
38				
39	ANNUAL SAVINGS BY USING CONTRACTOR / MA(			=(D34-D36)*D22
40				

Sheet1 / Sheet2 / Sheet3 /

APPENDIX C:
INSPECTING SECONDHAND FARM MACHINERY

INSPECTING A SECONDHAND TRACTOR

The overall appearance
An orderly inspection of the external appearance of a tractor can tell you if the machine has been cared for.

Silencer
Check for rust or bending.

Seat
Check for wear and heavy use.

Drawbar and rear linkage
Check all parts are fitted. Loose or sloppy action shows the need for repairs. A bent or misshapen drawbar indicates general tractor abuse.

pto
pto slipping under a heavy load needs adjusting. If adjusting does not correct slipping, then a new pto clutch needed.

Metal
Check radiator grille, diesel tank, and sheet metal parts for dents, rusted-out spots, or broken-off attaching clips. Damage/abuse can be seen.

Electrical system
Inspect the wiring for signs of fraying cables, broken lights etc.

Wheels
Inspect for front bearings wear which results in wheel wobble and for cracked rims.

A DETAILED INSPECTION OF THE DIESEL ENGINE

Oil mixed with water in the radiator or on the dipstick
Cracked block, cracked head, or leaky head gasket.

Rust streaks or dripping water around the water pump shaft
Water pump needs replacing or repairing.

Water showing on radiator core
Leaky radiator which can sometimes be repaired.

Rattling noise in fan
Bearing failure in the water pump.

Low oil pressure
Worn camshaft bearings, oil gauge broken, worn main bearings, worn oil pump.

Foam in the oil (water)
Cracked cylinder head or block or wet liners.

Blue smoke from exhaust
Usually engine is burning oil. Could also indicate that valve guides, rings, and sleeves are worn. Major engine overhaul may be needed.

Blowing excessive light smoke from the crankcase breather
Indicates engine is worn and needs overhauling.

Water seepage around head gasket or presence of foam in the radiator
Cracked head or head gasket damaged, maybe due to overheating.

Oil leakage around engine main bearings
May need new seals or bearings.

Seepage around a diesel injector
Repair or replacement of injector.

Diesel injection pump runs unevenly
Injection pump repair needed.

Engine missing while running
Valves or diesel injectors/pump needs attention.

The Tractor Chassis

Slipping clutch
Clutch needs overhauling or adjusting.

Clutch chatter
Worn clutch or possibly transmission bearings leaking oil onto the plate.

Rattle or scraping sound when you depress clutch
Thrust bearing worn.

No free travel of clutch pedal
Needs adjusting otherwise worn out.

Gear stick hard to move
No gear box oil, gears damaged, worn or seized.

Tractor jumps out of gear
Selector forks damaged.

Back axle whine or rumble
Damaged transmission.

Brakes inoperable
Seized or worn out and in need of replacing.

Finally, operate the tractor and check that gauges are working properly, the engine idles correctly and check the colour of smoke.

CARRYING OUT AN INSPECTION OF A SECONDHAND COMBINE HARVESTER

Elevators

Open clean grain and return elevator doors and inspect flights and chains for wear. Check main crop feed elevator casing. Ensure crop feed elevator bars are straight, and check connections (rivets/bolts) to the drive chain.

Drum and concave
Open inspection door above drum and concave, and inspect rasp bars and concave for wear and damage. Ensure rasp bars aren't bent (if steel) or cracked (if cast iron). Heavy vibrations during operations may indicate a missing or severely damaged rasp bar.

Straw walkers, grain pan and sieves
Inspect these internal components for signs of wear and damage.

Unloading auger
Inspect unloading auger in the grain tank carefully. Check the unloading auger and its tube.

Belts and pulleys
Check the condition of all drive belts, inspecting for chafing.

Operating test
Operate the combine and check the controls to find worn linkages, worn slip clutches, and other defects which show up only under actual field conditions.

Brakes
Test the brakes, ensuring they work evenly.

Transmission
Check that variable speed pulleys operate smoothly, that the combine picks up speed/slows down smoothly.

Table or header
Inspect, and note any damage to the reel tines; damage to the auger flights; worn, bent, or missing retractable fingers on the cross auger. Inspect the fingers and knife for damage.

Hydraulic system
Check the hydraulic system. If the header is raised and then settles by itself without you moving the actuating lever, the system is leaking or needs repair. Check oil levels; if low it may indicate lack of maintenance.

Inspect the engine
Check for oil and water leaks, and smooth running as in the description for tractors above.

APPENDIX D: EXAMPLE MAINTENANCE RECORDS

FIGURE D.1. TRACTOR LOG BOOK

Tractor model...Reg. no...

Serial nos. ....

Engine no. ....

Transmission no. ....

	Lubricating oil type	quantity	change interval	filter no.
Engine				
Transmission				
Hydraulics				
Power steering				

Other service requirements and intervals

Air filter ....

Fuel filter ....

Cab filter ....

Tyre pressures ....

Grease points ....

FIGURE D.2 DAILY RECORD OF USE AND SERVICE

Front of record card

Date	Hour meter reading	Fuel into tank	Lubricating oil used	Service work done	Work done by tractor	Driver's name

Reverse of record card

REPAIR & COST RECORD

Date	Parts used & cost	Labour & cost	Work by

147

FIGURE D.3. WORKSHOP JOB SHEET

Machine .... reg no. .... hourmeter reading
....

Date started .... date finished .... Mechanic
....

Work instructions carried out ....
....
....
....
....
....

Date	Parts used & part number	Part cost	Lubricating oil used	Other parts used	Time taken

FIGURE D.4 EXAMPLE MAINTENANCE CALENDAR BASED ON OPERATIONAL HOURS

Hours of Operation	HOUR METER READINGS																
10-hour service																	
10-hour service																	
10-hour service																	
10-hour service																	
10-hour service																	
50-hour service																	
100-hour service																	
250-hour service																	
500-hour service																	
1000-hour service																	

INDEX